Notified in A.C.Is. 5th August, 1944.

226
Publications
52

RESTRICTED

The information given in this document is not to be communicated, either directly or indirectly to the Press or to any person not authorized to receive it.

HANDBOOK OF ENEMY AMMUNITION

PAMPHLET No. 12

GERMAN GUN AND MORTAR AMMUNITION

By Command of the Army Council,

THE WAR OFFICE,
5th AUGUST, 1944.

LONDON:

GERMAN FUZE kl.A.Z.23

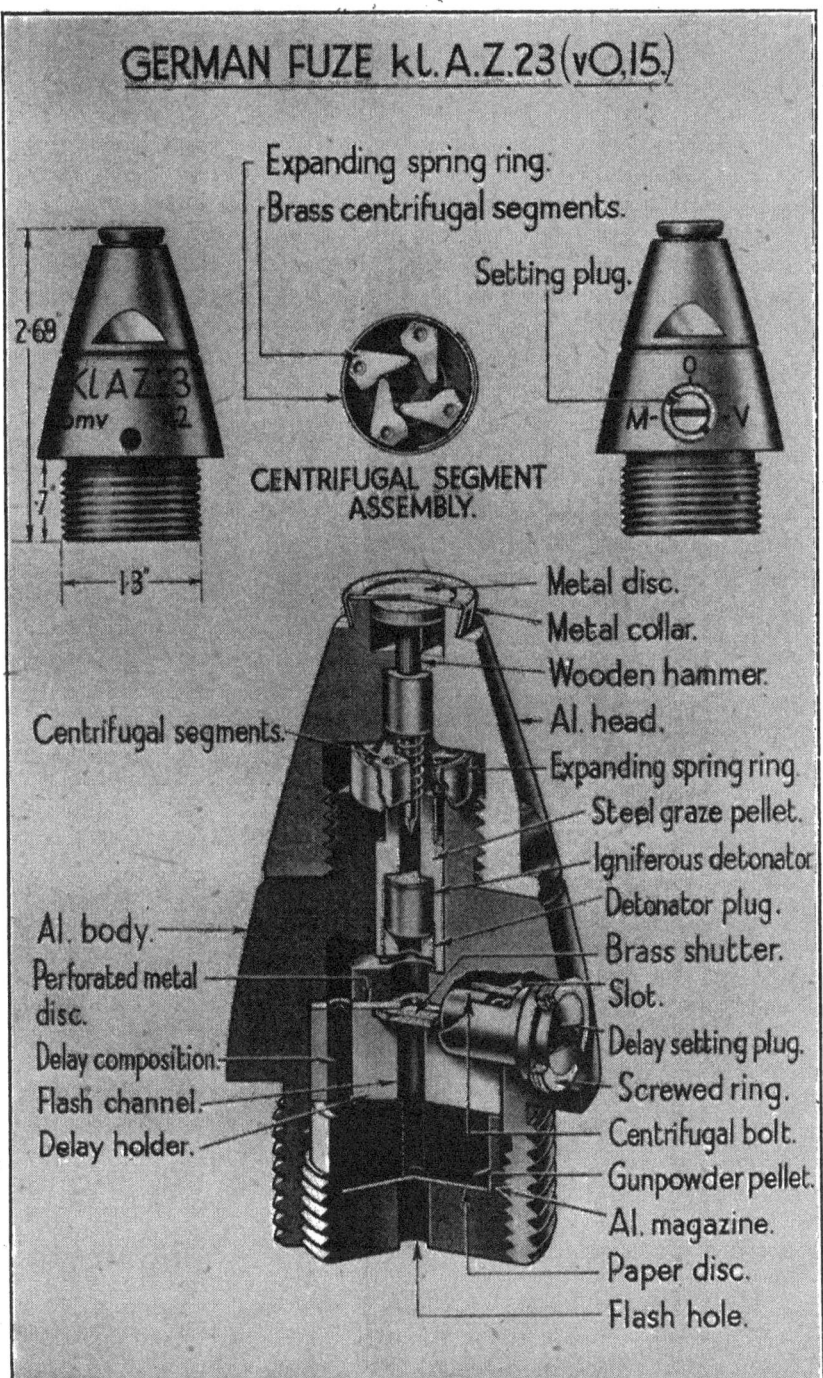

Fig. 1

HANDBOOK

OF

ENEMY AMMUNITION

CONTENTS TABLE

German Ammunition

	Page
Fuze kl.A.Z.23	1
Fuze kl.A.Z.23 Nb. Pr.	2
Fuze A.Z.23 v. (0,15)	4
Fuze A.Z.23/28 v. (0,1)	7
Fuze A.Z.23/42 v. (0,15)	9
Fuze A.Z.23 umg 0,15	9
Fuze Wgr. Z.36	13
Fuze, mechanical, time and percussion S/90K (Dopp Z.S/90K)	16
Base Fuze, Skoda B.Z.15 for 3·7 cm. Czech Shell	17
Base Fuze, Skoda B.Z.15 for 4·7 cm. Czech Shell	17
Base Fuze, Bd. Z.5127	20
Smoke Box No. 11 for H.E. shell	22
Primer, electric, Q.F. and Mortar Cartridges C/23	22
5 cm. Mortar Cartridge 39 with full charge	24
5 cm. Mortar H.E. Bomb 36 with fuze Wgr. Z.38	26
s.10 cm. K.18 and lg. 10 cm. K.T. H.E. Streamlined Shell Gr. 19	28
17 cm. K. Mrs. Laf. Q.F. Cartridge	30
20 cm. Light Spigot Mortar H.E. Bomb 40 and cartridge	35

ii

This combined graze and direct action fuze with an optional delay of 0·15 or 0·2 second is a smaller size of the normal A.Z.23 and is used in H.E. shell for the following equipments :—

    7·5 cm. Kw.K. (Tank gun).
    Stu. G. 7·5 cm. K (S.P. assault gun).
    7·5 cm. Geb. G.36 (Mountain gun).
    7·5 cm. Kw. K.40 (Tank gun).
    7·5 cm. Pak. 40 (Anti-tank gun).
    7·62 cm. Pak. 36 (Anti-tank gun).

The weight of the fuze is 4 oz. 14 drs. and its dimensions compare with the A.Z.23 as follows :—

Fuze	Protrusion	Intrusion	Diameter of screwthreads	Pitch of Threads
kl.A.Z.23	1·99 inches	0·7 inches	1·3 inches	1·5 mm.
A.Z.23	3·75 ,,	0·65 ,,	1·96 ,,	3 mm.

The fuze differs from the kl.A.Z.23 Nb mainly in the addition of the optional delay arrangement and the magazine which contains a perforated pellet of gunpowder weighing approximately 53 grains.

The construction and action of the fuze, as shown in the drawing is the same as that of the normal A.Z.23 fuze.

GERMAN FUZE kl. A.Z.23 Nb WITH ENCASED PLASTIC BODY (kl.A.Z.23 Nb. (Pr))

(Fig. 2)

The fuze, as indicated by the letters "Nb" in the abbreviated designation, is intended for use in smoke shell but has also been found in H.E. shell for the anti-tank equipment 7·5 cm. Pak 40. The deep olive green painting of the exterior distinguishes this fuze, largely of moulded plastic with a steel casing, from the aluminium fuze of the same designation which is described with the 7·5 cm. smoke shell in Pamphlet No. 4.

The weight of the fuze is 5 oz. 2 drs. The dimensions are shown on the drawing.

The fuze body consists of a thin steel ogival casing attached by spot welding to an inner cylindrical steel casing both of which cover moulded brown plastic. The ogival casing has a hole at the top for the wooden hammer. The hole is closed by a brass disc overlapped by the casing. The cylindrical casing is inserted from the base of the ogival portion and attached about half way up the interior. The lower part of the cylinder protrudes and is screwthreaded for insertion in the shell. The plastic interior is in two parts. The larger part, forming the centre of the body, is shaped to accommodate the graze and direct action mechanism and has an interior screw-thread in the lower portion to receive the moulded plastic base

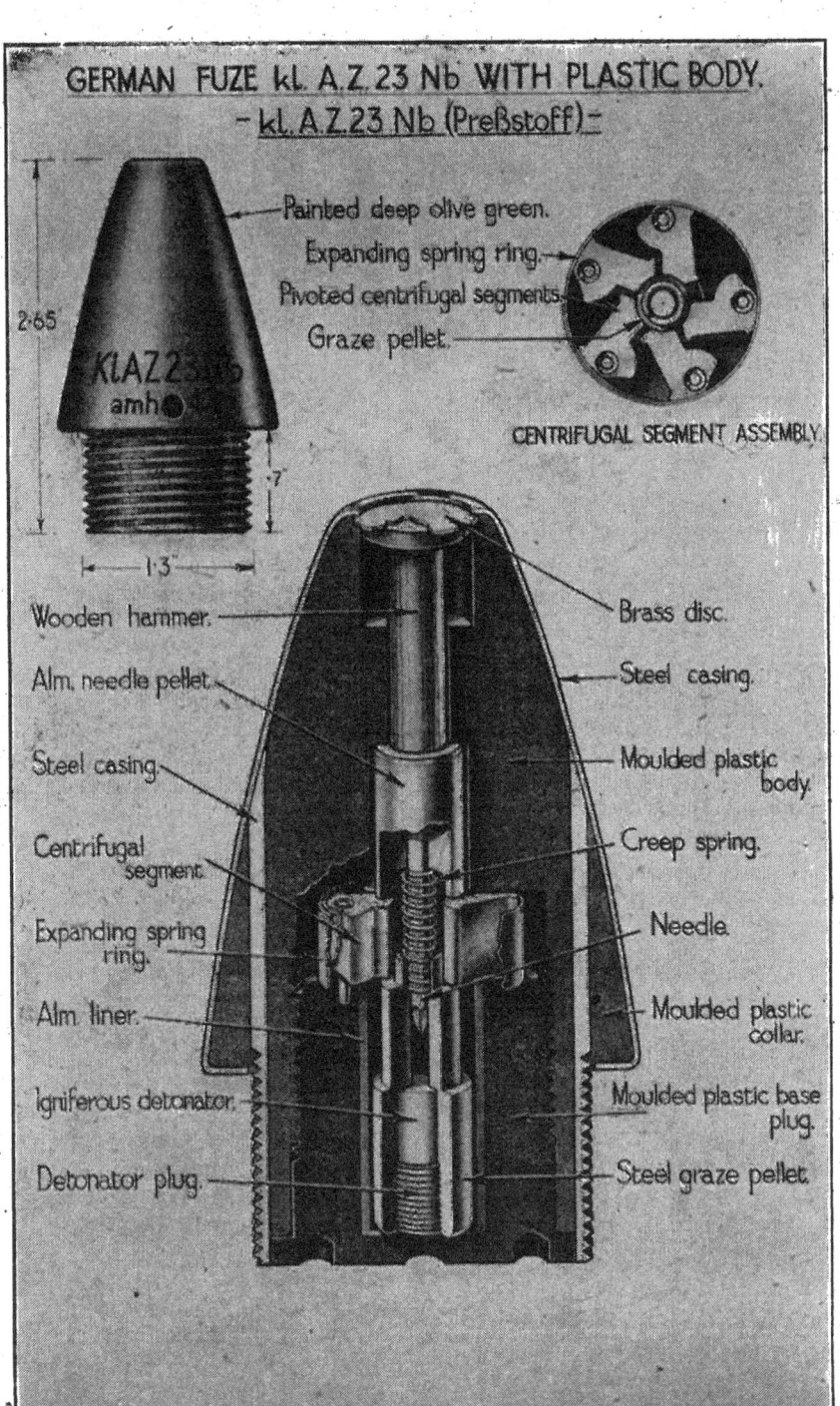

Fig. 2

plug. The smaller part of the plastic interior is in the form of a collar which surrounds the cylindrical casing within the ogival portion.

The fuze mechanism consists of a steel graze pellet carrying an igniferous detonator below a needle fitted to an aluminium pellet. The needle pellet is recessed at the base and is held away from the graze pellet by a creep spring and by five pivoted centrifugal segments. The segments are held in engagement with a shoulder on the graze pellet by a spring in the form of an expanding ring and are pivoted on the top of the plastic base plug. The plug is recessed and fitted with an aluminium liner to contain the graze pellet and has a flash hole at the base.

Action

The coil of the expanding spring ring is enlarged and the segments swing clear of the graze pellet by centrifugal force during flight. The needle pellet is then held in the forward position by the " creep " resulting from deceleration and is protected from air pressure by the brass sealing disc in the top of the fuze. Forward movement of the graze pellet is prevented by the creep spring.

On graze, the graze pellet sets forward, overcoming the creep spring, and impinges the detonator on the needle. With suitable impact the needle is simultaneously driven towards the graze pellet and direct action is obtained. The flash from the detonator passes through the flash channel in the detonator plug, and through the flash hole in the base of the fuze to the gaine beneath. The recess in the base of the needle pellet is apparently designed to fit over the top of the graze pellet and thus prevents the flash being expended in the wrong direction.

GERMAN FUZE A.Z.23v. (0,15)

(Fig. 3)

The fuze has a combined graze and direct action with an optional delay of 0·15 second and is used in H.E. shell of equipments as follows :—

Equipment	Shell
F.K. 16n.A. (7·5 cm.)	K.Gr. rot.
le.F.K.18 (7·5 cm.)	K.Gr. rot.
le.F.H.16 (10·5 cm.)	F.H.Gr.
	F.H.Gr.38 Stg.
le F.H.18 and le F.H.18M. (10·5 cm.)	F.H.Gr.
	F.H.Gr.38 Stg.
	F.H.Gr.F.
10 cm. K.17 and 10 cm. K.17/04 n.A. (10·5 cm.)	F.H.Gr. rot.
s.10 cm. K.18 (10·5 cm.)	10 cm. Gr.19.
m.10 cm. K.K. and m.10 cm. K.T. (10·5 cm.)	10 cm. Gr.34.
10·5 cm. L.G.40	F.H.Gr.41.
15 cm. K.18, 15 cm. K.39, 15 cm. K.(E) and 15 cm. K. Mrs. Laf.	15 cm. K.Gr. 18.

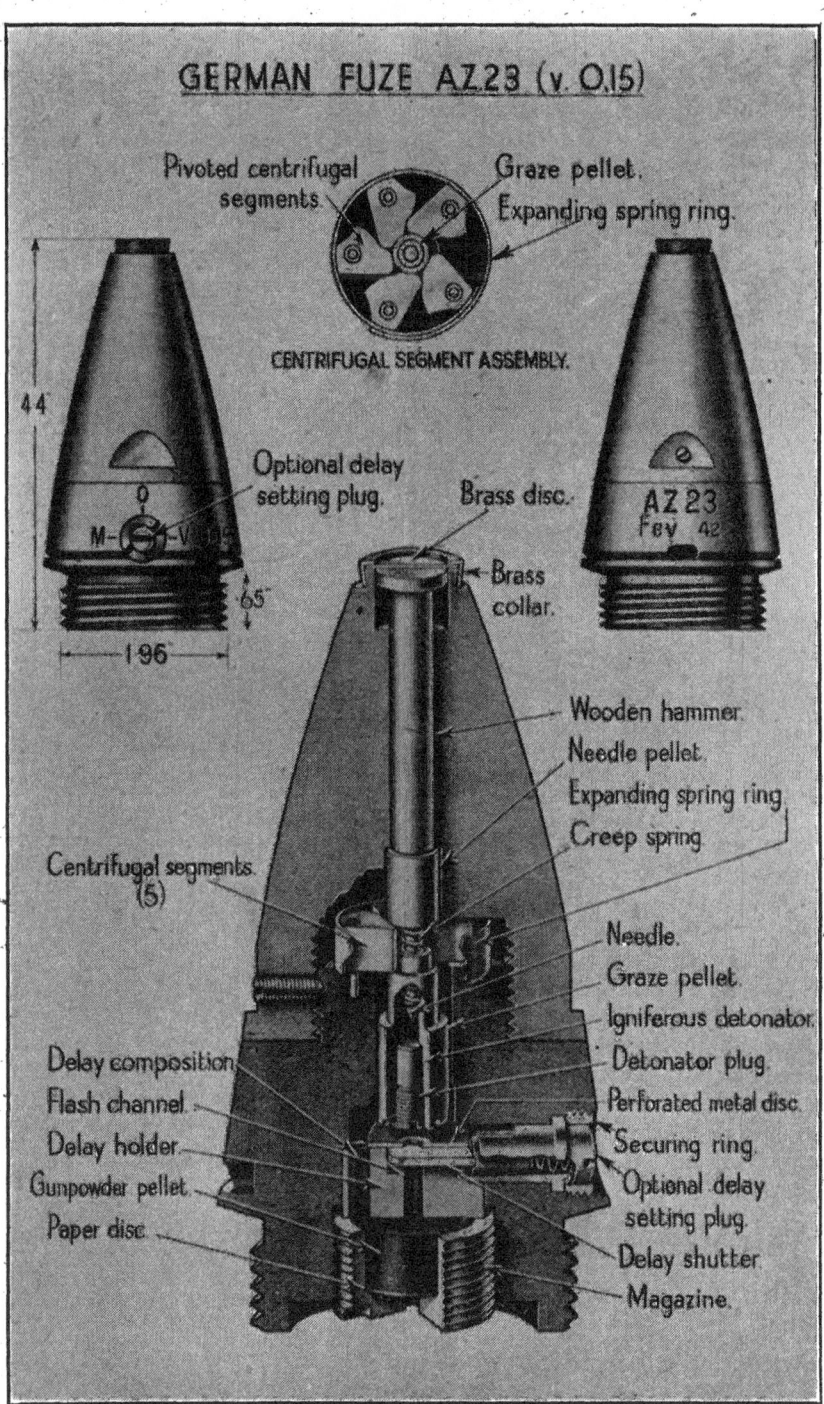

Fig. 3

The designation " A.Z.23 " is stamped above the flange of the aluminium body diametrically opposite to the optional delay setting plug. The period of delay is stamped adjacent to the plug in the form " V.0,15." To obtain delay the slot in the head of the setting plug is set coincident with the index marks lettered " M " and " V." For action without delay the plug is set to the " 0 " index.

The weight of the fuze is 15 oz. 6 drs. The dimensions are shown on the drawing.

The aluminium body of the fuze is in two parts. The head portion, which is screwed to the lower part, is solid and tapers towards the nose. A channel is formed through its centre to accommodate the wooden hammer and the needle pellet. The channel is closed against air pressure at the top by a brass disc which is secured by a brass collar fitted around a step formed in the nose of the fuze. The head is secured to the lower part by a fixing screw.

The lower portion of the body is tapered above the flange to correspond to the head and is screwthreaded below the flange for insertion in the shell. Near the top it is reduced in diameter and screwthreaded to receive the head and is recessed to accommodate the graze pellet. Another recess, formed in the base, contains the delay holder with a shutter and is screwthreaded to receive the magazine. The two recesses are connected by a central flash hole and an inclined flash channel. A radial channel for the optional delay assembly leads from the exterior to the lower recess.

The aluminium needle pellet, fitted with a steel needle, is supported above the graze pellet by a creep spring and by five centrifugal aluminium segments pivoted on the top of the lower portion of the body. The segments are held between the base of the needle pellet and a shoulder on the graze pellet by an expanding spring ring.

The graze steel pellet carries an igniferous detonator supported by a perforated screwed plug.

The delay holder consists of a cylindrical aluminium pellet with a flash channel through the centre and a second channel, displaced from the centre, which contains a delay filling and coincides with the inclined flash channel from the recess containing the graze pellet. A slot formed in the top of the holder to receive the shutter extends to just beyond the central flash channel. At the outer end of the slot the holder is recessed to receive the inner end of a centrifugal bolt forming part of the shutter. A tin disc with perforations corresponding to the channels in the holder is inserted above the holder.

The shutter assembly consists of a copper plate attached to a cylindrical bolt and is contained in the delay setting plug with a spiral spring which tends to retain the shutter in a position to close the central flash channel. The width of the shutter is greater than the diameter of the centrifugal bolt.

The delay setting plug is recessed from the inner end to accommodate the centrifugal bolt and the spiral spring and has two slots to receive the sides of the shutter projecting beyond the bolt. The outer end of the setting plug is closed and has a groove for the setting key. The plug is retained in the fuze body by a screwed securing ring which engages a flange on the plug but does not prevent it being turned in setting.

The magazine contains a pressed perforated pellet of gunpowder weighing approximately 2 drams and has a flash hole in the base closed by a paper disc.

Action

Before loading, the fuze is set for delayed or non-delayed action by means of the setting plug.

During flight the coil of the expanding spring ring is enlarged and the segments swung clear of the graze and needle pellets by centrifugal force. The needle pellet is then held in the forward position by " creep " whilst the graze pellet is held back by the creep spring. The action of the shutter is governed by the setting plug. With the plug set to " 0," the slots at its inner end are aligned with the projecting sides of the shutter and permit the centrifugal bolt to move outwards, taking with it the shutter and exposing the central flash channel in the delay holder. With the plug set in alignment with the " M " and " V " markings, the slots in its inner end are not in a position to receive the protruding sides of the shutter. The movement of the shutter and bolt is thus prevented and the shutter remains closed.

On graze, the graze pellet moves forward, compressing the creep spring, and impinges the detonator on the needle. With suitable impact, the hammer and striker pellet are driven in as the graze pellet moves forward and a more rapid action is obtained.

The flash from the detonator ignites the delay composition in the delay holder through the inclined flash channel and, if the shutter has opened, at the same time passes through the central flash hole and explodes the powder pellet in the magazine. With the fuze set for delay action, the central flash hole is masked by the shutter and the explosion of the magazine filling is brought about by the delay composition.

GERMAN FUZE A.Z.23/28 v.(0,1)

(Fig. 4)

This fuze is of the same construction, dimensions and weight as the normal A.Z.23 fuze with optional delay (*see* description of A.Z.23 v. (0,15) and Fig. 3 included in this pamphlet) but is fitted with a stronger spring between the graze and needle pellets instead

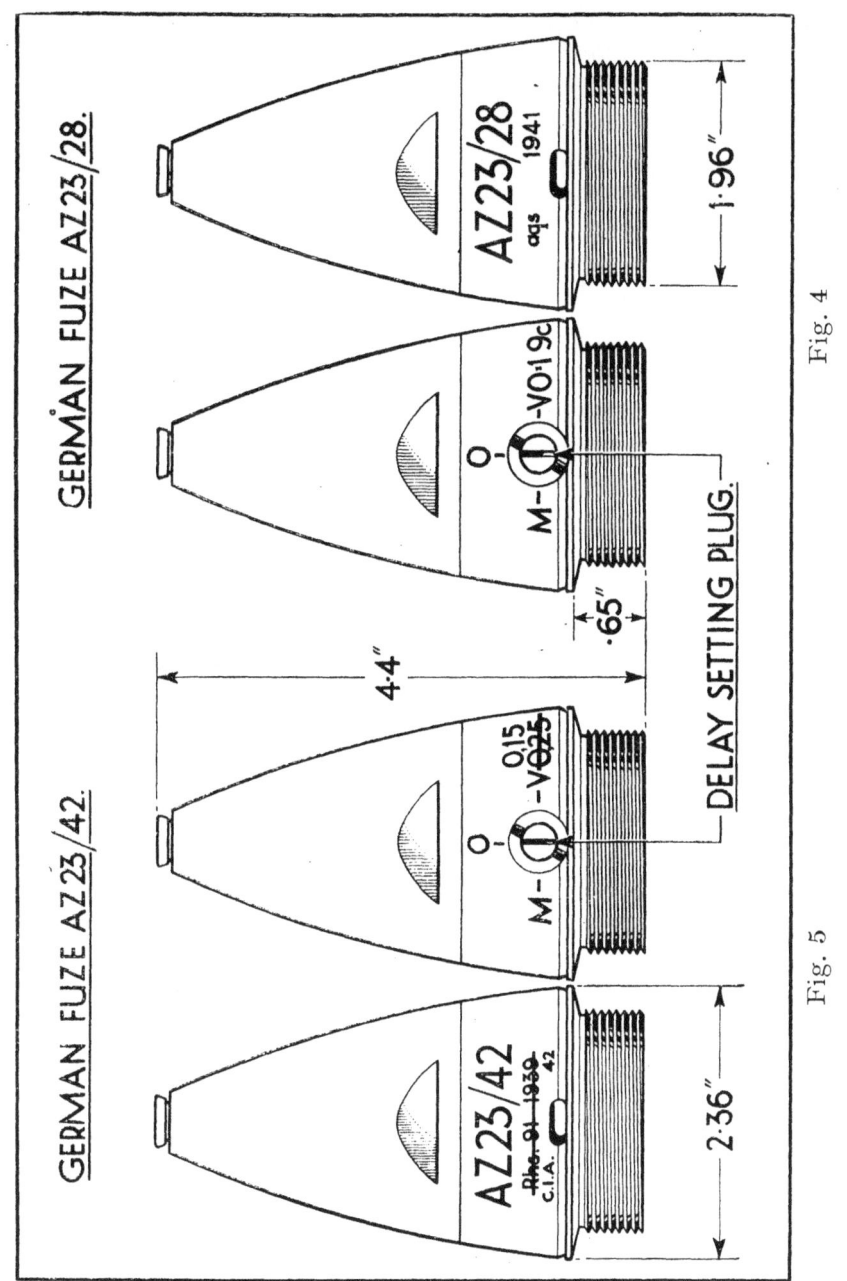

Fig. 4

Fig. 5

of the normal creep spring. The load required to bring the mechanism into the fired position is between 99 and 106 ounces as compared with a load of 16 to 19 ounces for the normal A.Z.23.

The fuze is identified by the designation " A.Z.23/28 " stamped above the flange where the time of the optional delay, 0·1 second, is also stamped.

The fuze is used in H.E. shell for the 8·8 cm. Flak 18, 36 and 41, (8·8 cm. multi-purpose guns), also for the 8·8 cm. Pak. 43 (anti-tank gun).

GERMAN FUZE A.Z.23/42 v. (0,15)

(Fig. 5)

This fuze is of the same construction, dimensions and weight as the normal A.Z.23 fuze with optional delay (*see* description of A.Z.23 v. (0,15) and Fig. 3 included in this pamphlet) but the centrifugal segments and the expanding spring ring encircling them are designed to arm at a lower rotational speed. The fuze is armed by centrifugal force at rotational speeds between 3,000 and 4,200 r.p.m. and the A.Z.23 v. (0,15) between 4,500 and 5,500 r.p.m.

The fuze is identified by the designation " A.Z.23/42 " stamped above the flange. The fuze examined had been stamped to indicate a delay of 0·25 second but this had been barred out and 0·15 stamped above. It is used with the H.E. shell for the 10·5 cm. Geb.H.40 (mountain howitzer).

According to a German document the fuze can also be used instead of the A.Z.23 v. (0,15) in every type of shell for the 10·5 cm. l.F.H.18 in which the latter is used.

GERMAN FUZE A.Z.23 umg 0,15

(Fig. 6)

The A.Z.23 umg fuzes differ externally from the A.Z.23 fuzes in having a much shorter coned head and are longer between the flange and the base. This lower part, which enters the fuze hole in the shell, has only a few screwthreads below the flange, the remainder being plain. Three types of the " umg " fuze have been found. These are :—

A.Z.23 umg 0,8.
A.Z.23 umg m.2V.
A.Z.23 umg 0,15.

The " 0,8 " fuze with an optional delay of 0·8 of a second, and originally used for 15 cm. Gr.19 for the 15 cm. medium howitzer s.F.H.18, is described in Pamphlet No. 1. This description was based on a French report. Fuzes of this type of later manufacture are made of steel with a rustproofed surface.

The " m.2V " fuze is also an optional delay fuze with two alternate delays. Details of the fuze are not yet available. Both this and the " 0·8 " fuze are being replaced by the " 0,15 " fuze.

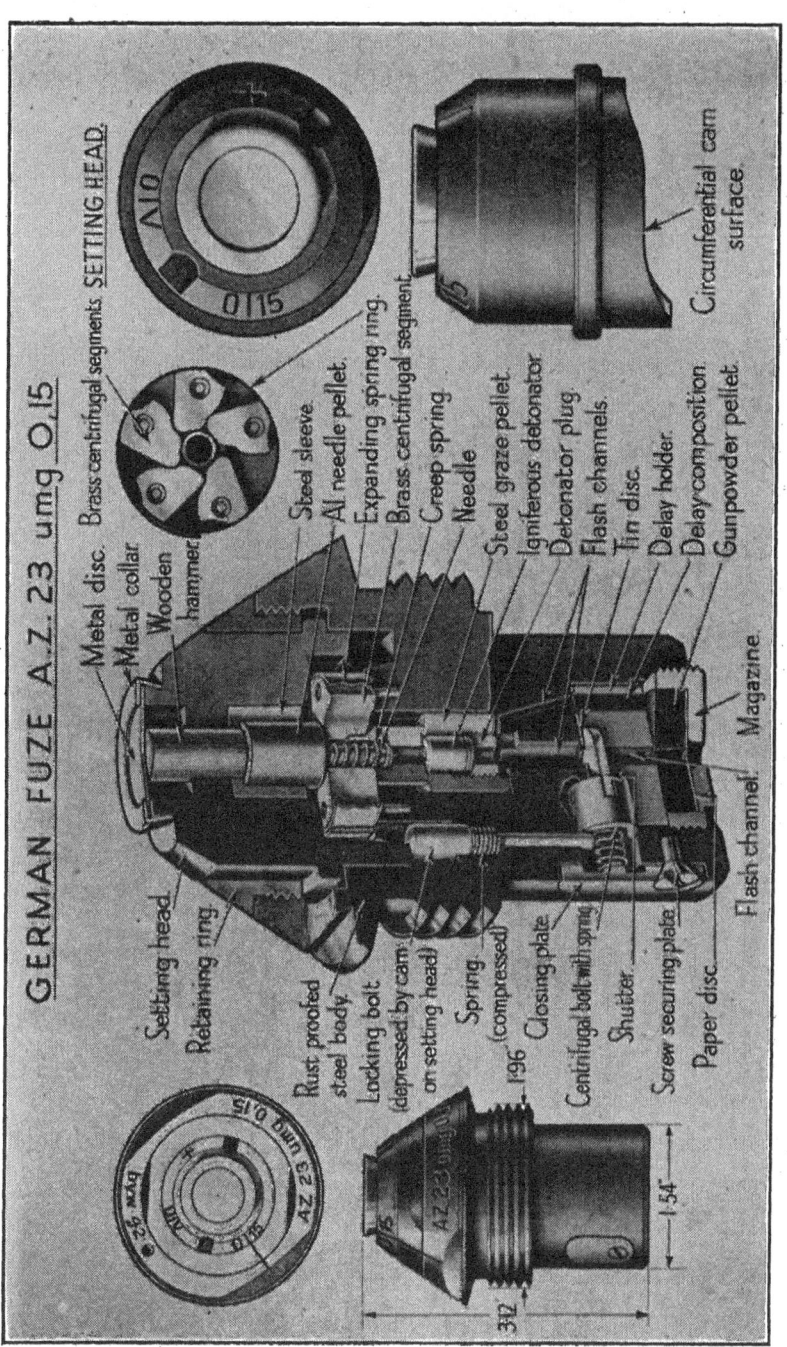

Fig. 6

The " 0,15 " fuze with an optional delay of 0·15 of a second is here described and differs from the later type of " 0,8 " fuze only in the time of the delay.

Each of these fuzes can be identified by the designation stamped above the flange. The " m.2V " and more recently the " 0·15 " fuze are used in the H.E. shell, " 15 cm. Gr.19 " for the 15 cm. medium howitzer s.F.H.18 and in the H.E. shell " 21 cm. Gr.18 " for the 21 cm. howitzer 21 cm. Mrs. 18. The weight of fuzes of this type is approximately 1 lb. 10 oz. 2 drs.

The A.Z. umg 0,15 has the usual combined D.A. and graze mechanism consisting of a steel graze pellet carrying the detonator and an aluminium needle pellet held apart by a creep spring and five brass centrifugal segments encircled by an expanding spring ring. The graze pellet is contained in a central recess in the body which has two flash channels at the base. One channel leads direct to the magazine through an open channel in the delay holder, the other leads to the magazine through the delay composition in the delay holder. The needle pellet with a steel sleeve surrounding it and a wooden hammer above it is carried in the head of the fuze. The steel head is retained in the body by a retaining ring which screws into the top of the body and engages an external flange near the lower part of the head. The circumferential rim at the bottom of the head varies in depth to act as a cam which positions the spring loaded locking bolt of the shutter. Thus when the head is rotated to bring the deepest portion of the rim over the locking bolt the bolt is pressed down into the radial recess containing the shutter in the lower part of the fuze and prevents the shutter from opening. The setting positions of the head are marked on its protruding part by two lines at right angles which are set to an index line on the retaining ring and body. The setting line for delayed action is marked " MV " and that for " non-delay " " OV."

The delay holder, shutter assembly and magazine are all of the same type described for other fuzes of the A.Z.23 type.

Action

The fuze is set for " delay " or " non-delay " by turning the head to bring the appropriate setting marking into coincidence with the index line. When set to " MV " the locking bolt is pressed down by the deep part of the rim at the bottom of the head and locks the shutter in the closed position. When set to " OV " the short part of the rim is over the locking bolt thus permitting the bolt to be raised clear of the shutter recess by its spring.

During flight the coil of the expanding spring ring is enlarged and the segments swung clear of the graze pellet by centrifugal force. If set for " non-delay " the shutter is also thrown outwards and the open flash channel in the delay holder is exposed. Forward movement of the graze pellet is prevented by the creep spring.

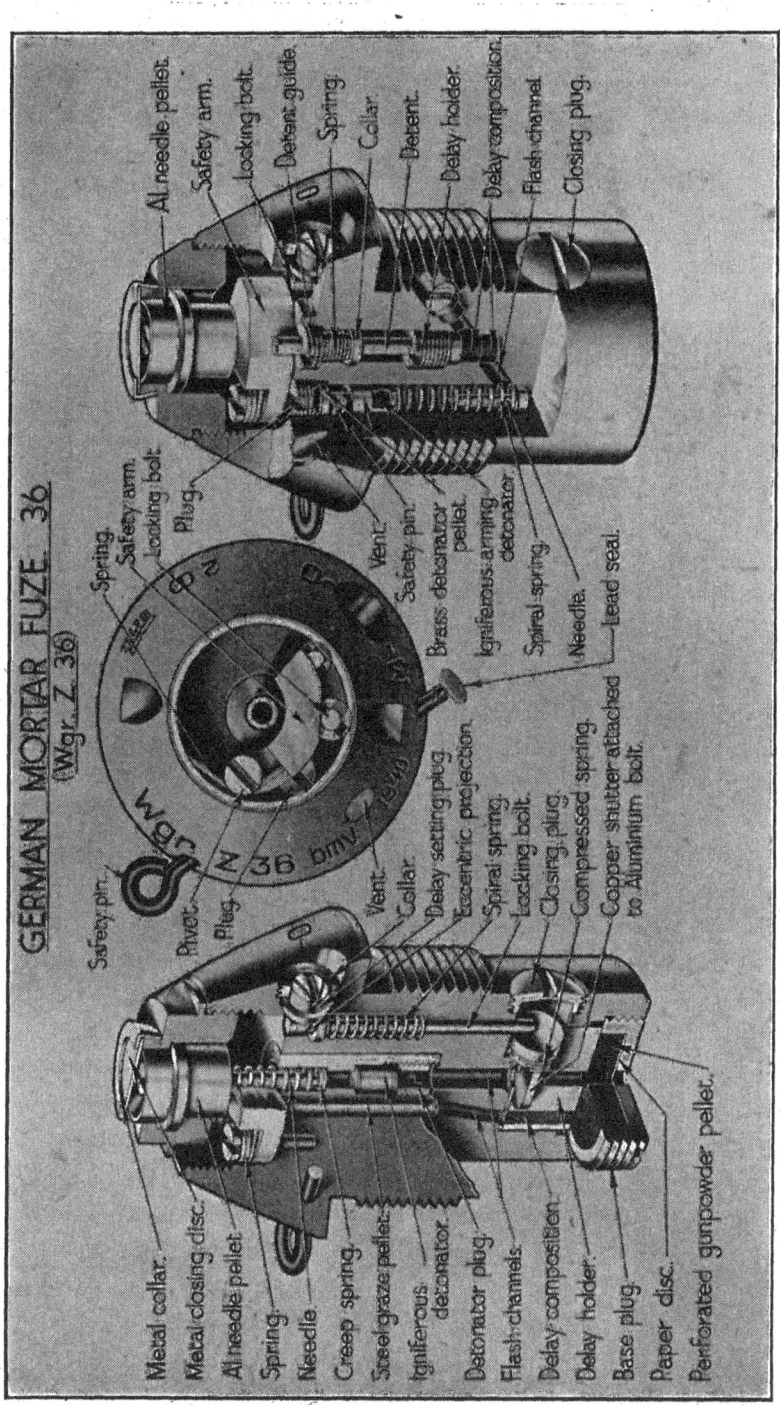

Fig. 7

On graze the detonator is carried forward by the graze pellet and pierced by the needle. When suitable impact is obtained the hammer and needle pellet are driven in as the graze pellet moves forward, thus accelerating the action. With the head set for "non-delay" the flash from the detonator passes direct to the magazine. If set for "delay" the open channel in the delay holder is masked by the shutter so the flash has to burn through the delay composition to reach the magazine.

GERMAN FUZE Wgr. Z.36

(Fig. 7)

This igniferous nose fuze, used with a gaine in the H.E. bomb fired from the 20 cm. light spigot mortar, is designed to arm in an unrotated projectile and has a combined graze and direct action with an optional delay. The arming device is operated by the burning of a delay composition which is ignited by a second detonator and delays the arming for a period of approximately 0·6 of a second after firing.

The body of the fuze is of aluminium alloy with a flat topped conical head which has the usual metal sealing disc at the top and is fitted with a safety pin secured by a lead seal. The abbreviated designation "Wgr. Z.36" is stamped near the lower part of the coned head. On the opposite side of the head there is a delay setting plug inscribed with an arrowhead. Stamped in the head at one side of the plug is a graduation lettered "M." On the opposite side a similar graduation is lettered "O." The arrowhead is set to the "M" graduation for delay or the "O" graduation for non-delay action.

The weight of the fuze is 6 oz.

The fuze body is recessed at the top to receive the needle pellet and safety arm and screwthreaded internally for the insertion of the closing cap. At the base of this recess four vertical recesses are formed. The central and largest recess contains the steel graze pellet and has two flash channels at the base. One of these channels is central, the other is displaced from the centre and is inclined. Of the other vertical recesses, two are formed side by side and are connected near the base ends by an inclined flash channel, which continues to the exterior of the body where it emerges at the screwthreaded portion and is closed by a screwed plug. One of these recesses contains a brass detonator pellet carrying a No. 37 igniferous detonator, the pellet being supported above a steel needle by a spiral spring and a safety pin. The pin is inserted through the head of the fuze and engages in a circumferential groove in the side of the detonator pellet. This recess is closed at the top by a screwed plug. The connected vertical recess contains, at its lower end, two small

pellets of delay composition beneath a sleeve filled with a similar composition. An inclined channel leads from the base of the recess to the head of the fuze where it is lightly closed by a metal disc secured by stabbing. The sleeve is of aluminium or aluminium alloy and is screwthreaded externally for insertion in the recess. A detent positioned above the sleeve is supported by the delay filling contained in the sleeve. The detent has a collar formed around its centre which supports a spiral spring held under compression between the collar and a screwed ring at the top of the recess. The upper end of the detent passes through the screwed ring and retains the safety arm in the safe position, i.e. partly covering the central recess which contains the graze pellet and closing the fourth recess. The fourth vertical recess has a channel at its base leading to a radial channel containing the shutter bolt in the lower part of the fuze. This recess contains a locking bolt for the shutter. The bolt is supported by a spring and has a collar formed near its head. The collar is engaged by an eccentric projection on the inner end of the delay setting plug which is contained in a radial channel in the head of the fuze. When the plug is set to the " M " graduation the eccentric projection, bearing down on the collar, pushes the locking bolt down into the shutter bolt channel and prevents the bolt moving outwards. When the plug is set to " O," the eccentric projection is raised and the locking bolt is free to be raised clear of the shutter bolt when the safety arm moves clear of the recess. The safety arm, in the form of a curved aluminium arm, is pivoted at one end where a spring is fitted which tends to swing it clear of the path of the graze and needle pellets and clear of the recesses containing the locking bolt.

The lower part of the fuze contains the shutter assembly, the delay holder and a magazine of gunpowder. The shutter assembly consists of a copper shutter attached to an aluminium bolt and a spiral spring which is fitted to the inner end of bolt and is under compression when the shutter is held in the closed position by the locking bolt. The delay holder is of the usual type used in German fuze and is in the form of a solid aluminium cylinder with a central flash channel with a groove for the shutter above it and a displaced channel containing delay composition. A perforated tin disc is placed over the holder and the shutter. The magazine screwed into the base of the fuze is in the form of cup shaped closing plug with a central flash hole and contains a perforated pellet of gunpowder. The flash hole is closed by a paper disc on the inner side.

Action

The safety pin is removed and the fuze set for delayed or non-delayed action before firing. If set for " delay " the locking bolt is held down to engage the shutter bolt by the eccentric projection on the setting plug. If set for " non-delay " the locking bolt is held down only by the safety arm.

On acceleration, the detonator pellet in the side recess sets back, compressing its spring, and the detonator is pierced by the needle. The flash passes through the connecting channel into the recess containing the delay composition and the detent. The pressure set up by the burning of the delay composition escapes by the inclined channel leading to the exterior of the head of the fuze. When the delay composition supporting the spring loaded detent has been destroyed, the detent is forced down by its spring and the safety arm is thus released. The arm is then swung clear of the graze and needle pellets and the recess containing the locking bolt by its pivot spring. If set for " non-delay " the locking pellet is then free to be raised by its spring and the shutter is released to be pushed out by its compressed spring to the open position. If set for " delay " the locking bolt cannot rise as it is held by the projection on the setting plug and the shutter remains closed leaving only the delay channel in the holder exposed. Forward movement of the graze pellet during the period of deceleration is prevented by the creep spring.

On graze the pellet overcomes the spring and carries the detonator on to the needle. When suitable impact is obtained, the needle pellet is driven in at the same time as the pellet moves forward and direct action results. The path of the flash from the detonator to the magazine is governed by the setting. If set for " delay " the central channel in the delay holder is closed by the shutter and the flash can pass only through the channel containing the delay composition. If set for " non-delay " the central channel is exposed and the flash passing by this route will be the first to reach the magazine.

Detonator and Delay Compositions

The detonator carried in the graze pellet is No. 26 which is in common use in German fuzes. The No. 37 detonator contained in the side recess has a layer consisting of mercury fulminate, potassium chlorate and antimony sulphide over a filling of black powder.

The delay compositions in the sleeve supporting the detonator and the pellets beneath the sleeve consist of the following :—

	Sleeve	Pellet
Nitrocellulose	3·9 per cent.	2·7 per cent.
Red lead	75·5 ,, ,,	72·0 ,, ,,
Silicon	20·6 ,, ,,	25·3 ,, ,,

The delay composition used in the channel in the delay holder has been found to be gunpowder in a fuze dated 1940. In a fuze dated 1941 a composition similar to that in the sleeve and pellets has been found. This composition has also been found in the delay holder of the A.Z.35K fuze described in Pamphlet No. 11.

GERMAN MECHANICAL TIME AND PERCUSSION FUZE S/90 K. (Dopp. Z.S/90 K)

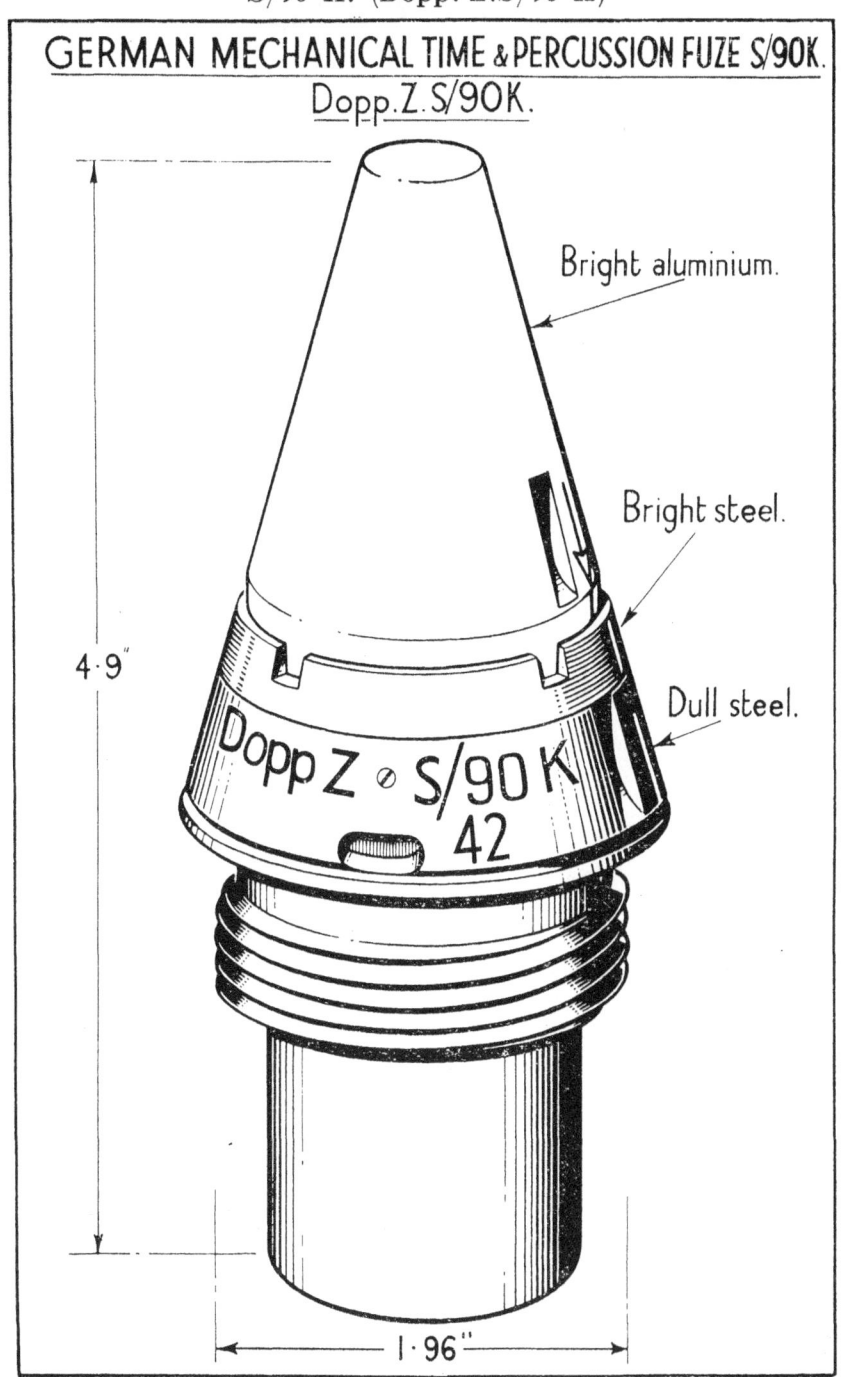

Fig. 8

The fuze has a mechanical time action with a maximum time of running of 90 seconds and a graze action of the normal German type. The construction and action are the same as described and illustrated for the Dopp. Z.S/90/45 in Pamphlet No. 11.

The fuze is used in the H.E. shell (K.Gr.39) and the H.E.B.C. shell (K.Gr.38 (Hb)) for the 17 cm. gun (17 cm. K. Mrs. Laf.).

The weight of the fuze is approximately 1 lb. 12 oz.

GERMAN BASE FUZE, SKODA B.Z.15

(Fig. 9)

Details of the filling of a gaine fitted to this fuze for use in the 3·7 cm. A.P.C. shell of Czech origin are shown on the drawing.

The fuze is described in Pamphlet No. 2, pages 6 and 7, and the shell in Pamphlet No. 9, page 19.

GERMAN BASE FUZE, SKODA 15 FOR 4.7 cm. Pak (t) AND Pak. K.36 (t) A.P.C. SHELL

(Fig. 10)

The fuze is used in the 4·7 cm. A.P.C. shell of Czech origin described in Pamphlet No. 10, page 29, and differs from that described for the 3·7 cm. A.P.C. shell in Pamphlet No. 2, pages 6 and 7, mainly in the addition of a delay sealing plug.

The steel needle pellet and creep spring are contained with the detonator in a cylindrical holder which is screwed into the base of the steel head of the fuze. The head contains the delay arrangement and, at its forward end, carries the gaine.

The delay arrangement consists of a sealing plug which closes the flash hole leading to the gaine and a filling of gunpowder which surrounds the plug. The cylindrical plug has a pintle and copper washer at the top to seal the flash hole and a flange near the base. The clearance between the flange and the wall of the recess in which the plug is contained is 0·015 inch. Below the flange there are four radial channels leading to a recess in the base of the plug.

Action

During flight the needle pellet, with the two locking balls, moves forward as the result of deceleration, until the balls are in a position to be thrown outwards by centrifugal force. The needle is then held off the detonator only by the creep spring.

On impact, the needle pierces the detonator and, at the same time, the sealing plug sets forward and seals the flash hole leading to the gaine. The flash from the detonator enters the recess in the base of the plug, and passing through the radial channels, ignites the

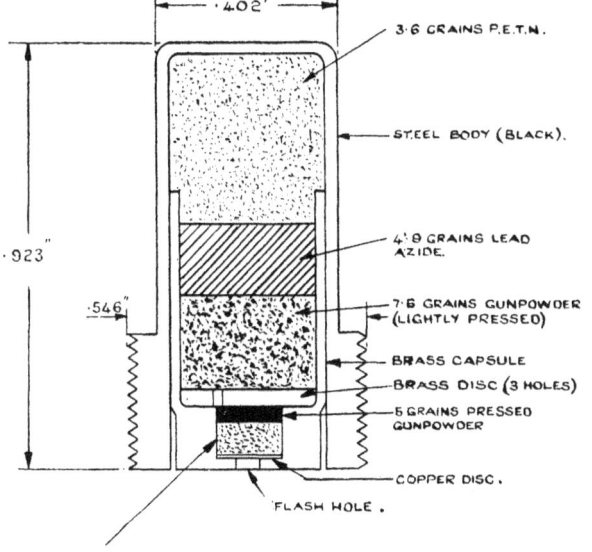

Fig. 9

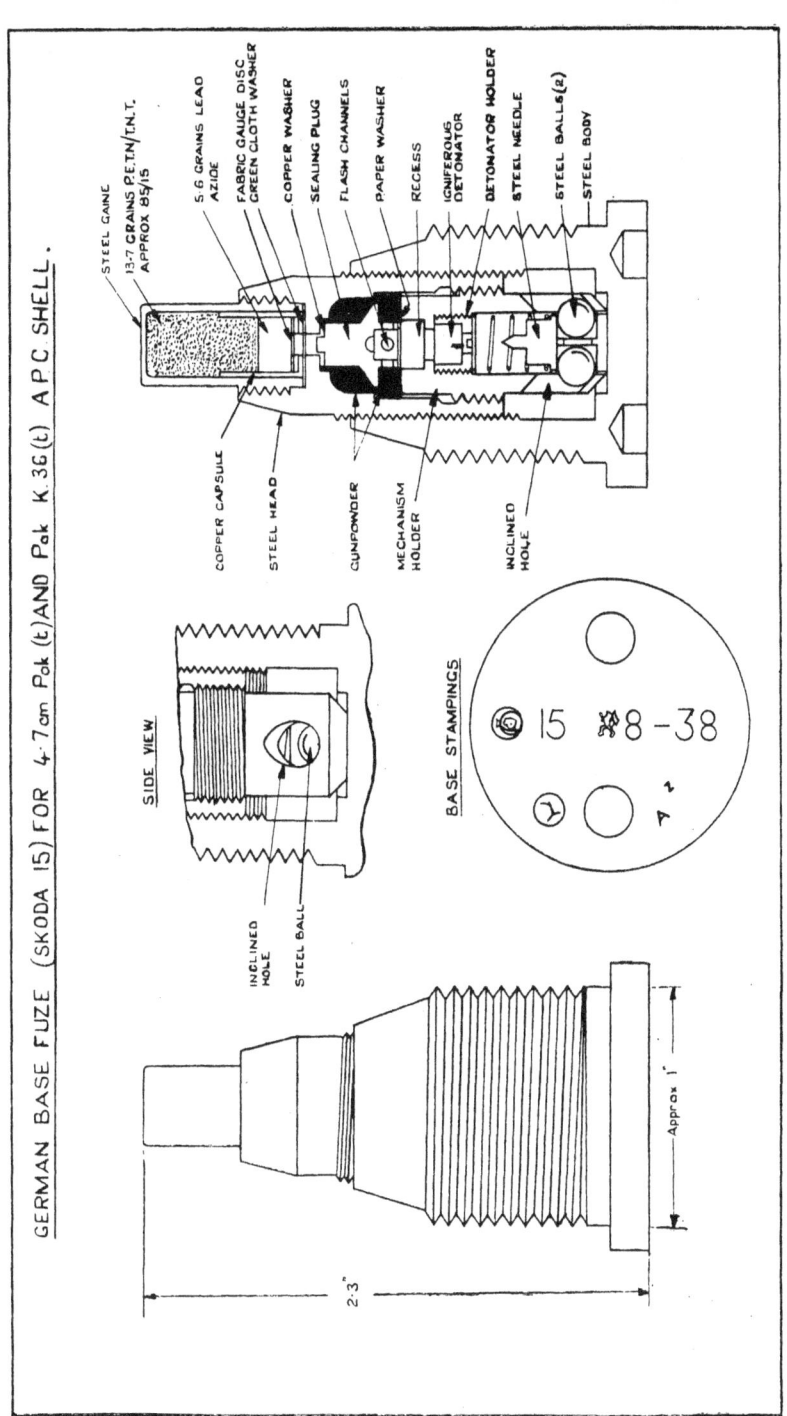

Fig. 10

gunpowder. The pressure set up by the ignition of the gunpowder behind the flange presses the plug firmly into the flash hole above. The combustion of the gunpowder leaves the pellet unsupported and leads to the ignition of the gunpowder in front of the flange, through the small clearance between the flange and the wall of the recess. The resultant pressure acting on the front of the flange pushes the sealing plug back into the recess in the forward end of the mechanism holder and thus opens the flash hole leading to the gaine.

GERMAN BASE FUZE Bd. Z.5127
(Fig. 11)

This is a base fuze fitted with a tracer and is used in the A.P.C.B.C. shell for the 8·8 cm. Flak 41 multiple purpose gun and for the 8·8 cm. Pak 43 anti-tank gun. The weight of the fuze with tracer is 11¾-oz. The tracer alone weighs ¾ oz. The fuze body is approximately 2·3 inches long, and screwthreaded externally at one end for insertion in the shell. The interior is divided into two compartments by a diaphragm formed in the body. The rear compartment contains the tracer and is rather smaller than that in the front which is screwthreaded internally and contains the fuze mechanism.

The fuze mechanism consists mainly of a tubular mechanism holder having at its rear end a fixed pellet containing a detonator and at its front end a striker with compressed spiral spring, two steel balls and an inertia collar.

The tubular holder is screwthreaded externally for approximately two thirds of its length from the base, its forward end is slightly less in diameter and is surrounded by a loose fitting inertia collar secured by a shear wire. The base of the holder is closed by a cup-shaped pellet containing an igniferous detonator. Nearer the forward end of the holder are two radial channels diametrically opposite and closed on the outside by the inertia collar.

Two steel balls are held partially in the channels and partially in two recesses in the body of the striker thereby holding the latter off the detonator. One end of the striker spring, which is held under compression, bears against a shoulder in the channel at the forward end of the holder whilst the other end fits into the cup-shaped body of the striker. Two flash holes are formed in the base of the striker body on diametrically opposite sides of the needle.

The front end of the fuze body is closed by a gaine containing cyclonite over lead azide and lead styphnate.

Action

On impact, the inertia collar sets forward and breaks the shear wire thereby allowing the steel balls, under centrifugal action, to move outwards and unlock the striker. The striker under the action of its spring is forced back on to the detonator. The flash from the

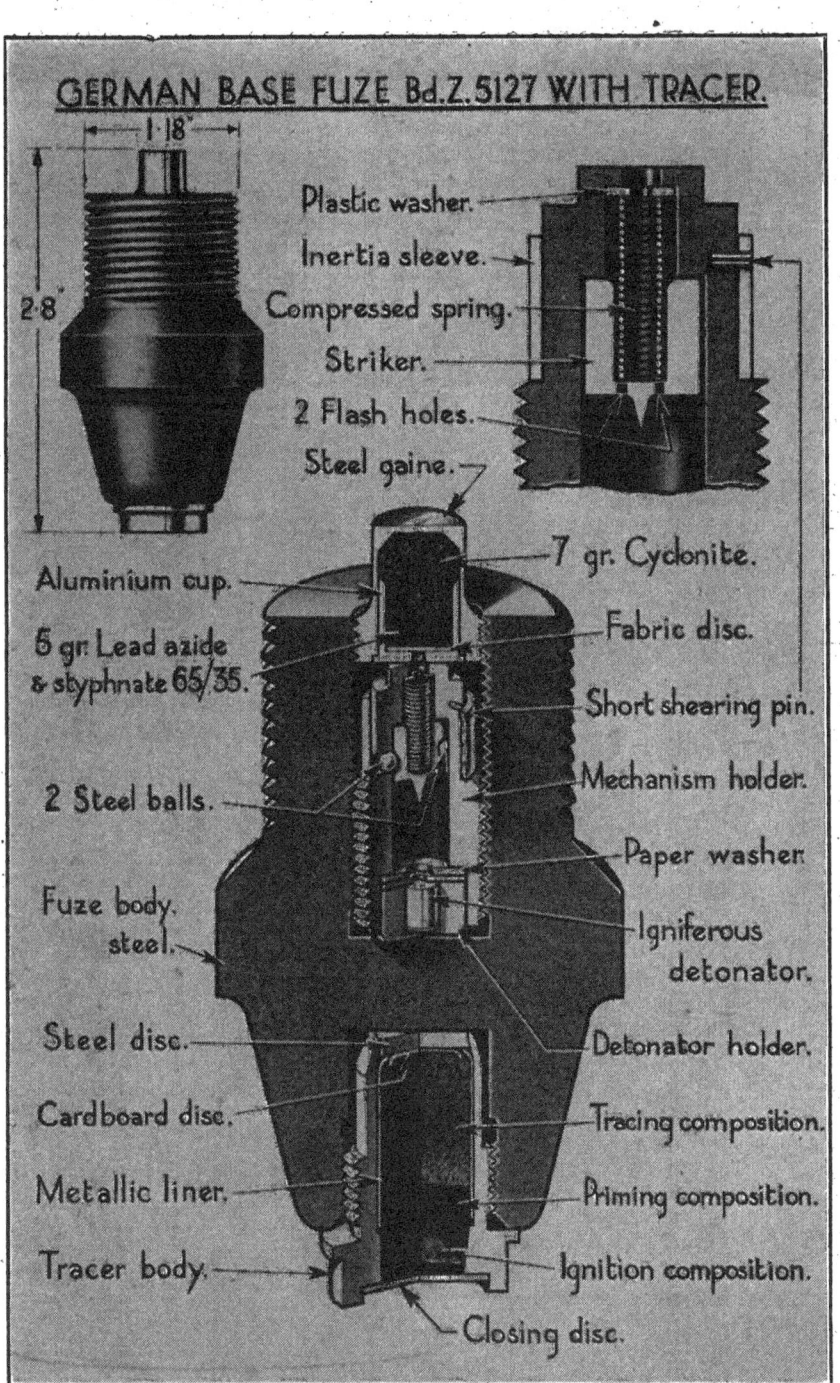

Fig. 11

detonator passes through the flash channels on either side of the striker to the gaine at the forward end of the fuze.

GERMAN SMOKE BOX NO. 11 FOR H.E. SHELL
(Rauchentwickler Nr. 11)

The box consists of a bakelized cardboard cylindrical container, 4 inches long and 1 inch in diameter, closed at the bottom with a yellow disc of heavily bakelized paper. The top is closed with a red plastic cap fitted inside the container. The composition weighs 76·05 grams (approximately 2 oz. 11 drs.) and consists of :—

Red phosphorus	85·4 per cent.
Paraffin wax	11·6 per cent.
Magnesium phosphates	2·9 per cent.

The smoke box examined had the marking " 3 KOLA 3976 " arranged in three lines on the top and " KOLA " at the base.

GERMAN, PRIMER, ELECTRIC, Q.F. AND MORTAR CARTRIDGES, C/23
(Fig. 12)

This is a smaller model of the C/22 electric primer described in Pamphlet No. 4, page 31, and is used in the cartridge for the 20 cm. light spigot mortar and in the fixed Q.F. cartridge for the 3·7 cm. Kw.K. (Tank gun). The primer corresponds to the percussion primer C/13 in dimensions and is identified by the designation " C/23 " stamped in the base. The contact plug is visible in a chamfered hole in the base.

The brass body has two key flats formed at the base and is screwthreaded for insertion in the cartridge. The diameter over the threads is ·52 inch and the threads 18 to the inch.

The brass contact plug, contained inside the lower part of the body, is cylindrical in shape with a stem at the bottom which emerges through a chamfered hole in the base of the body for contact with the firing mechanism. A small lug is formed near the top of the plug which fits into a recess in the body to prevent rotation. A layer of light brown insulating composition coats the side wall of the plug including the stem and the greater part of the top. The top of the plug is recessed to contain gunpowder and the electric fuze head which extends radially across the plug. The fuze head consists of an upper and a lower contact strip with insulating material between them and a bridge wire at one end surrounded by a blob of ignition composition. The lower strip is in contact with the top of the contact plug whilst the upper strip is in contact with a projection inside the brass contact washer assembled above, but insulated from, the plug. The washer, in addition to this projection, has two external projections or lugs which engage in recesses inside the body

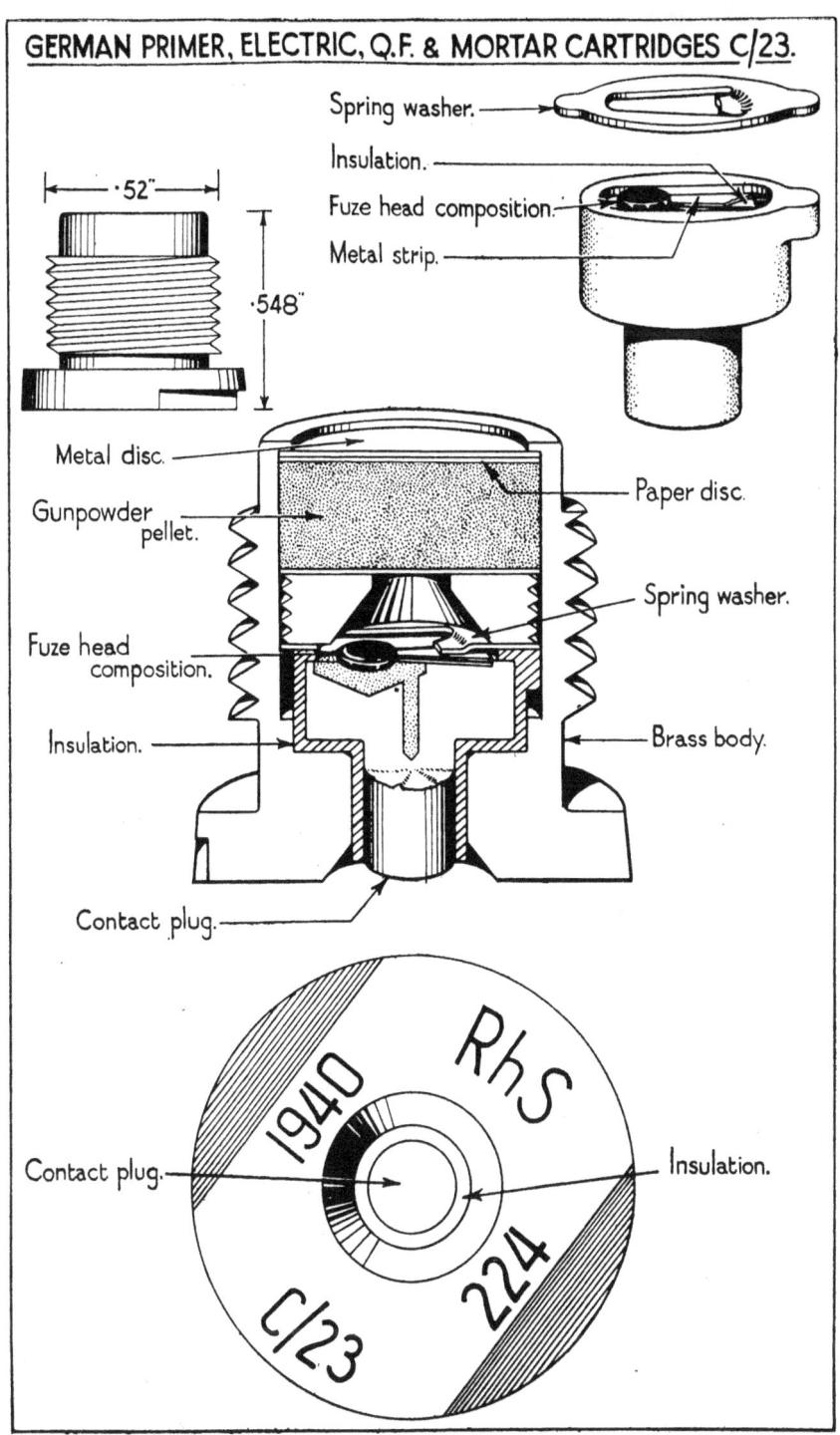

Fig. 12

and prevent rotation. A screwed plug with a central flash hole and a concave underside is screwed into the body to secure the contact washer and cover a priming of gunpowder surrounding the fuze head. The magazine in the top of the primer contains a pellet of gunpowder and is closed by a thin metal disc. The weight of the pellet is approximately 7 grains.

The path of the firing current is from the firing mechanism, through the contact plug to the lower strip of the fuze head, through the bridge wire to the upper strip thus heating the wire and igniting tne composition and thence by the contact washer, through its inner projecting piece, through the body to earth.

GERMAN 5 cm. MORTAR CARTRIDGE 39 WITH FULL CHARGE—5 cm. Wgr. Patr.39 (gr. Ldg)
(Fig. 13)

The cartridge is used with the H.E. bomb (5 cm. Wgr.36) described in Pamphlet No. 4 and is identified by the marking " 5 cm. 39 " at the base. The base is lacquered green to indicate the full charge. A similar cartridge with a reduced charge, the " 5 cm. Wgr. Patr 39 (kl. Ldg) " has a red base.

The cartridge is used without augmenting cartridges and is of the usual primary type for mortars. The rolled paper body is green and the typical 28 bore sporting cartridge base is brass plated. The dimensions are shown on the drawing.

Body

The rolled paper body is lacquered externally and is fitted with a lining tube also of rolled paper. The body is closed at the mouth by a cardboard wad which is held by the rim being turned inwards. At the base the body is strengthened by a steel liner in the form of a cup fitting externally over the end. The steel liner is covered by a copper liner which extends further up the body and is in turn covered by a brass-coated copper shell which forms the base. A wad of rolled black paper holds the cap chamber in the body within the metal liners.

Cap

The cap chamber is of steel with a coating of copper and contains a brass cap in its lower part. The cap contains a brass anvil and a 0·73 grain filling consisting of :—35 per cent. of lead styphnate, 4 per cent. of tetrazene, 43 per cent. of barium nitrate, 6 per cent. of antimony sulphide and 12 per cent. of calcium silicide. The cap rests on a soft copper foil disc which closes the base of the cap chamber. The base is lacquered green.

Propellant Charge

The propellant charge of nitrocellulose powder consists of a priming charge of cylindrical grains in the lower part of the body and

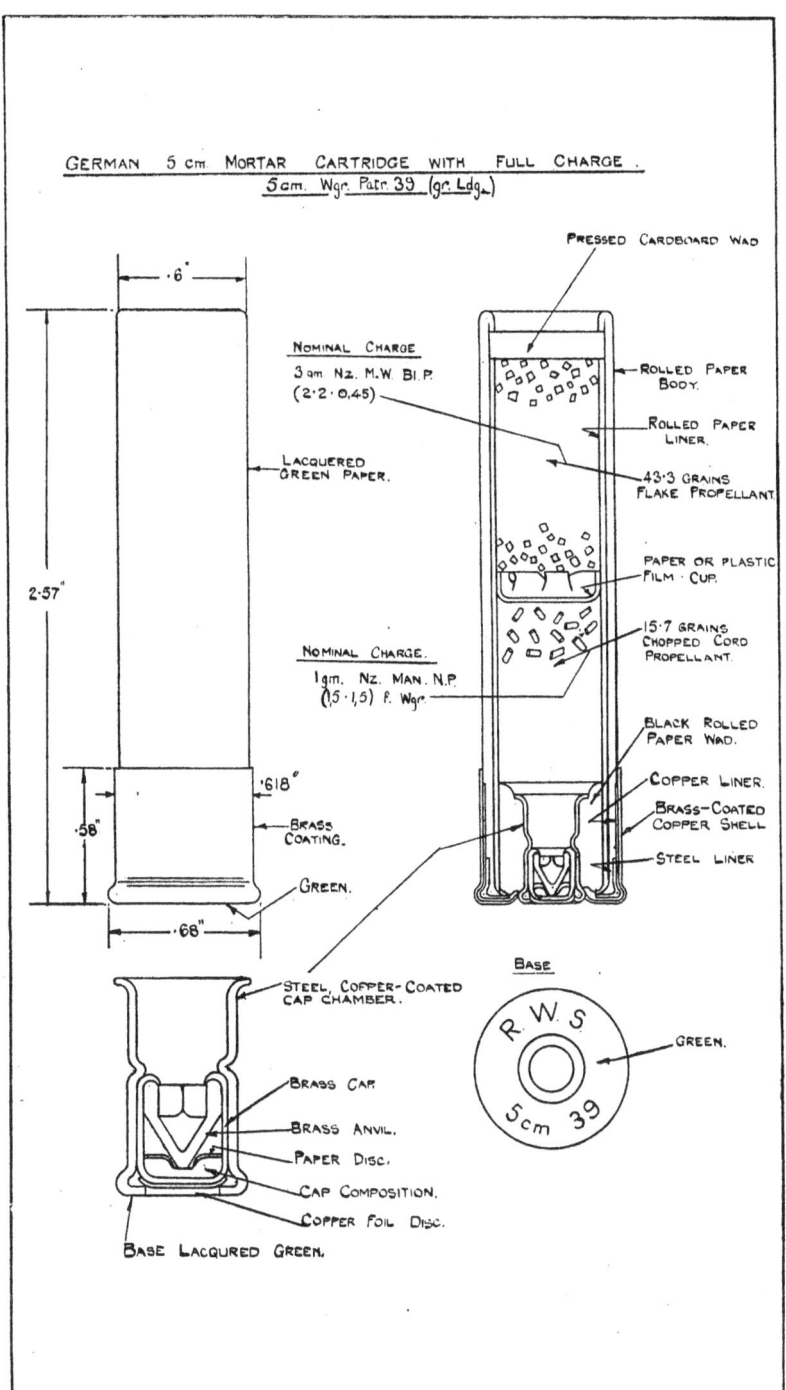

Fig. 13

a main charge of a square flake in the upper part. The two charges are separated by a paper or plastic film cup.

The priming charge has a nominal weight of 1 gram and the propellant is designated "Nz.Man.N.P.(1,5·1,5) f.Wgr.". In the cartridge examined the charge weighed 15·7 grains. The propellant, in the form of pale grey-green, porous chopped cords, with graphite incorporated, has the following composition :—Nitrocellulose plus graphite 95·3 per cent., diphenylamine 1 per cent., potassium sulphate 1·1 per cent. and included 2·6 per cent. of volatile matter. The nitrogen content of the nitrocellulose is 13·1 per cent.

The main charge has a nominal weight of 3 grams and the propellant is designated "Nz.M.W.B1.P.(2.2.0,45)." In the cartridge examined, the charge weighed 43·3 grains. The square flake propellant, lightly coated with graphite, has the following composition :—Nitrocellulose (nitrogen content 13 per cent.) 98·1 per cent., diphenylamine 0·6 per cent. and included 1·3 per cent. of volatile matter.

Both propellants burn at a faster rate than that of Ballistite B.16 and the cartridge should, therefore, be efficient under wet weather conditions.

GERMAN 5 cm. MORTAR H.E. BOMB 36 WITH FUZE Wgr. Z.38. (5 cm. Wgr. 36 m. Wgr. Z.38)

(Fig. 14)

The following details are additional to those given in Pamphlet No. 4 :—

Bomb

The lower portion of the cavity for the bursting charge, where the tail unit is screwed into the body, is sealed by a coating of bitumen followed by a pad of magnesium oxychloride cement.

Bursting Charge

The T.N.T. bursting charge is a cast filling with a density of 1·60 and a setting point of 80·4 degrees centigrade. The weight of the charge in a bomb recently examined was 3·5 oz.

Gaine

The gaine, carried in an aluminium exploder container which is screwed to the lower portion of the fuze, is a Kl.Zdlg.34 Np. This small gaine, filled P.E.T.N./Wax, is described in Pamphlet No. 11.

Fuze

The igniferous detonator, fitted in the base of the fuze, consists of a cylindrical copper shell closed at the head and base by a thin copper disc and containing a detonator composition over a 0·06 grain filling of gunpowder. The detonator composition weighs ·3 grains and consists of :—potassium chlorate 51 per cent., antimony sulphide 24·1 per cent. and calcium silicide and glass 24·9 per cent.

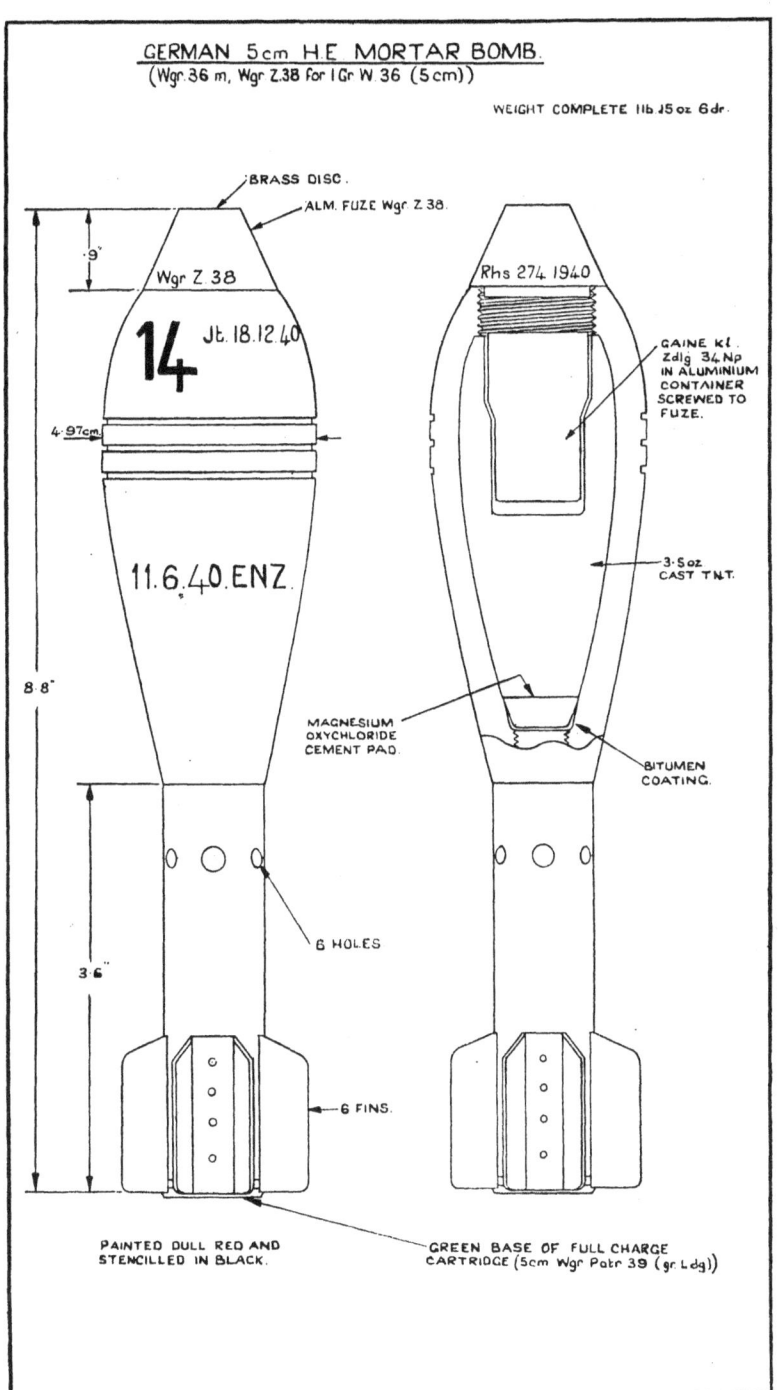

Fig. 14

GERMAN 10·5 cm., H.E. STREAMLINED SHELL FUZED A.Z.23 OR DOPP. Z.S/60s. (10 cm. Gr. 19)

(Fig. 15)

The shell is used in the 10·5 cm. medium gun " s.10 cm. K.18 " with the mechanical time and percussion fuze " Dopp. Z.S/60s " (described in Pamphlet No. 10) or with the combined D.A. and graze fuzes, with optional delay, " A.Z.23 v. (0,25) " or " A.Z.23 v. (0,15) " The fuze with 0·25 second delay is described in Pamphlet No. 1. A description of the fuze with 0·15 second delay is included in this pamphlet. The shell is also used in the long 10·5 cm. turret gun " lg. 10 cm. K.T." with the " Dopp.Z.S/60s " fuze.

The body of the shell is painted the normal deep olive green, is stencilled in black and has two driving bands. The stencilling includes the H.E. numeral near the nose (" 13 " indicating amatol 40/60 and " 14," cast T.N.T.), the weight class in Roman numerals at the shoulder and the smoke box marking below the shoulder (R.11 indicates the inclusion of smoke box No. 11). The smoke box marking is omitted in some instances although this component is present.

The fuzed shell is 19·1 inches in length and weighs 33 lb. 9 oz. 8 dr. when filled and fuzed. Each of the fuzes used has a protrusion of approximately 3·7 inches.

Shell

The shell is in two parts, the head being screwed into the body about half way up the ogive. The body is of forged steel and is fitted with two copper clad driving bands of iron. The cavity in the upper part of the body is cylindrical. In the lower part it tapers towards the base and is machined. The head is a machined forging and is screwthreaded internally at the nose for the insertion of an adapter fitted with an exploder container and for the fuze. The exploder container is of mild steel. The weight of the empty shell is 26 lb. 2 oz. 6 drs. The diameter at the shoulder is 4·11 inches (10·44 cm.).

Method of Filling

The bursting charge consists of cast T.N.T. or amatol 40/60 with a cavity below the fuze hole which contains a No. 11 smoke box beneath the exploder container. The weight of the amatol 40/60 bursting charge in a shell examined was found to be 3 lb. 12 oz. 15 drs.

The smoke box, Rauchentwickler Nr. 11 is described as a separate item in this pamphlet.

The gaine in the exploder container is the larger size of the C/98 model with a filling of P.E.T.N./Wax (Gr. Zdlg. 3/98 Np). Details of the gaine are included in Pamphlet No. 6.

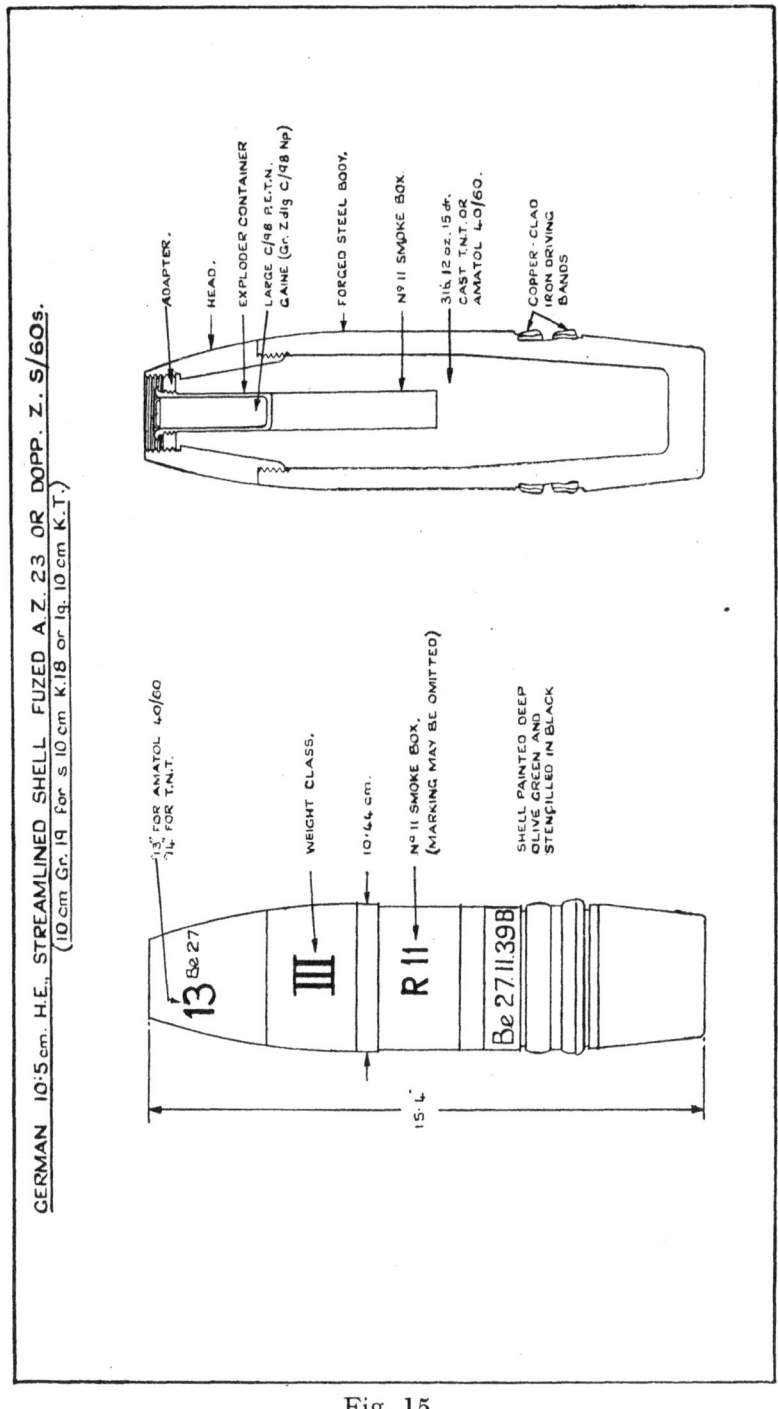

Fig. 15

GERMAN 17 cm. K. Mrs. Laf. Q.F. CARTRIDGE

(Figs. 16 and 17)

The cartridge is used in the 17 cm. K. in Mrs. Laf. (17·25 cm. gun mounted on the 21 cm. high angle semi-mobile carriage) and consists of the case with the percussion primer of the C/12 type and the propellant charge in five sections. Only one of the sections (the Haupkart) is contained in the case when packed, the remainder being packed in metal cylinders. The sections provide four charges but are marked with abbreviated designations instead of the usual charge section numerals. These markings are also found on the packages. The designations used are as follows :—

    Section 1. Sonderkart. 1.
    Section 2. Sonderkart. 2.
    Section 3. Haupkart.
    Section 4. Vorkart. 3.
    Section 5. Vorkart. 4.

The combinations of sections to provide the four charges are as follows :—

    Charge 1. Sonderkart. 1 in the case alone.
    Charge 2. Sonderkart. 1 with Sonderkart 2 extending down one side. Both sections in the case.
    Charge 3. Vorkart. 3 loaded into the front of the chamber and followed by the Haupkart contained in the case.
    Charge 4. Vorkart 3 with Vorkart 4, contained in its central tube, loaded into the front of the chamber and followed by the Haupkart contained in the case.

Charges 1 to 3 (inclusive) are used with the H.E. streamlined shell " Gr.39 " fuzed with the A.Z.35K or Dopp. Z.S/90K fuzes. Charge 4 is used with the H.E.B.C. streamlined shell " Gr.38 (Hb) " fuzed with the Hb.gr.Z.35K or the Dopp. Z.S/90K fuzes.

The case is stamped at the base with the model number 6342 and the designation of the equipment, " 17 cm. K.Mrs.L."

Propellant Charge

The propellant is of the Digl. double base type consisting basically of diethylene glycoldinitrate and nitrocellulose. The weight, nature and size used in the sections, as indicated by the markings, are :—

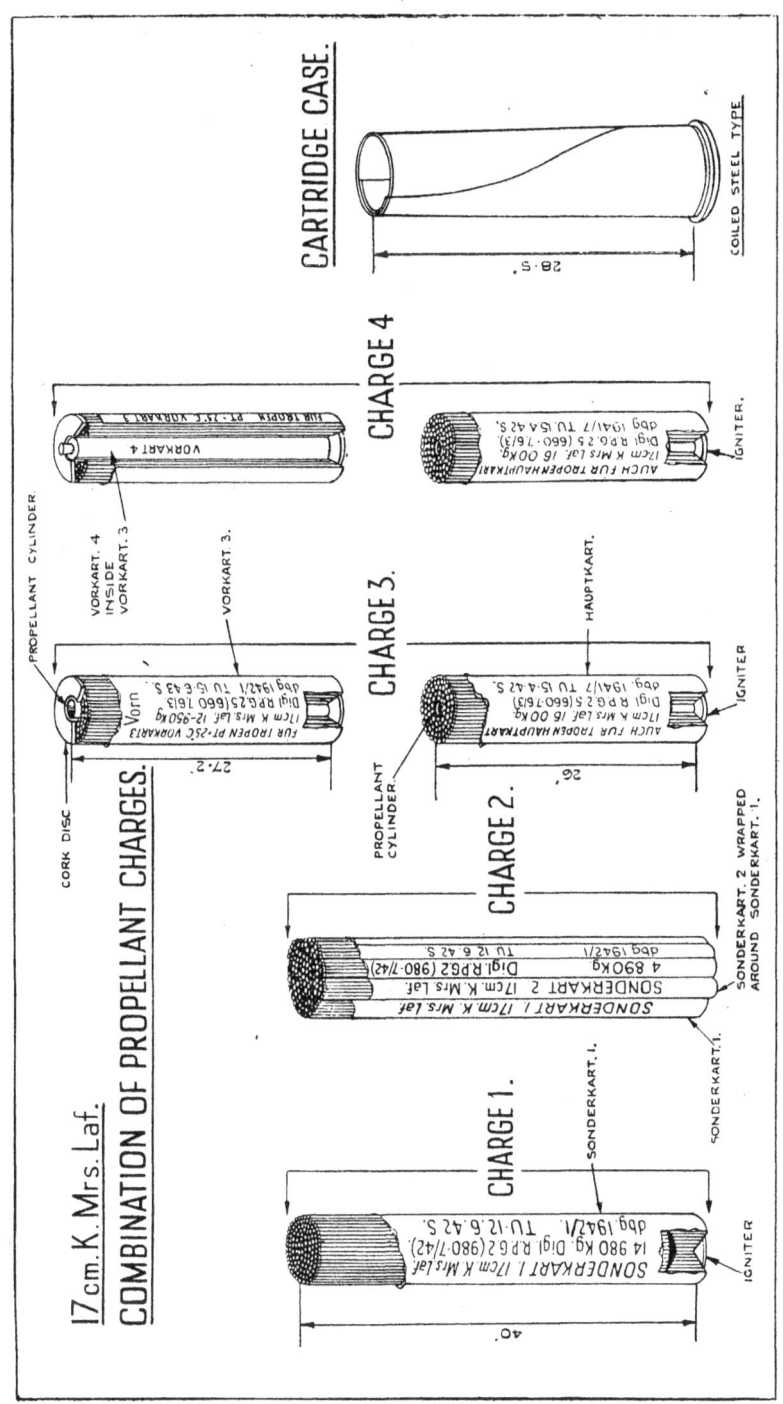

Fig. 16

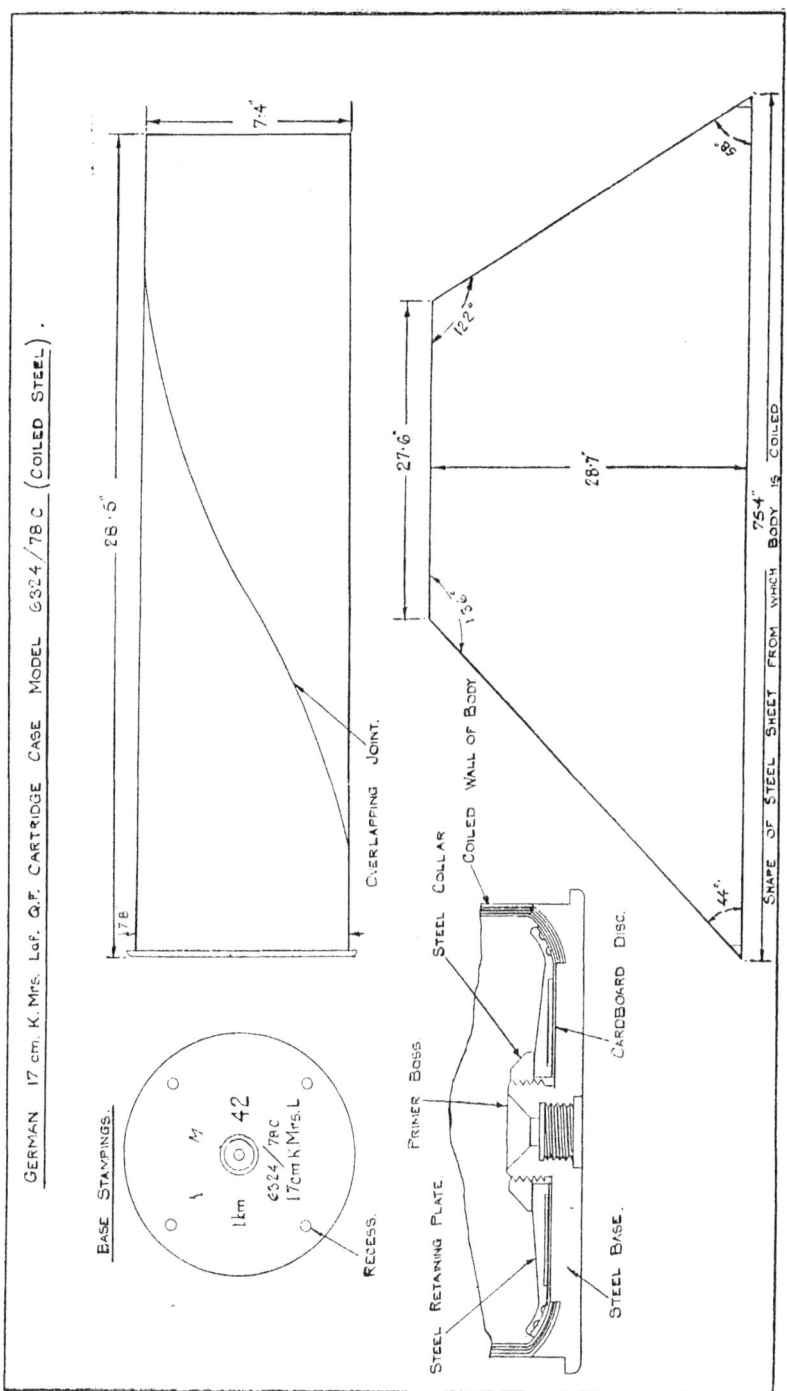

Fig. 17

Section	Weight Kilograms	Nature, shape and size	Weight lb. oz. dr.	Size in Inches			
				Length	External Dia.	Internal Dia.
1 (Sonderkart 1)	14·980 Kg.	Digl.R.P.–G2–(980–7/4,2)	33 0 5	38·6	0·276	0·165
2 (Sonderkart 2)	4·890 Kg.	Digl.R.P.–G2–(980–7/4,2)	10 12 8	38·6	0·276	0·165
3 (Haupkart)	16 Kg.	Digl.R.P.–G2,5–(660–7,6/3)	35 4 8	26	0·299	0·118
4 (Vorkart 3)	12·950 Kg.	Digl.R.P.–G2,5–(660–7,6/3)	28 8 14	26	0·299	0·118
5 (Vorkart 4)	1·350 Kg.	Digl.R.P.G2,5–(660–7, 6/3)	2 15 10	26	0·299	0·118

The weights of the propellant charges, as stencilled on the bags, vary considerably and are apparently adjusted charge weights based on the performance of the propellant lot at proof.

Construction of Charge Sections

Section 1 (Sonderkart 1) is approximately 40 inches in length and consists of a bundle of tubular cords of propellant contained in a white cylindrical bag. The bag is choked at the front end and carries an igniter at the base. The igniter contains approximately 100 grams of Nz. Man. N.P.(1,5–1,5). This is the nitrocellulose powder, in the form of cylindrical grains, normally used in German igniters.

Section 2 (Sonderkart 2) is approximately 40 inches in length and consists of five comparatively small bundles of tubular cord propellant contained in separate pockets formed in a white rectangular bag. The bag is sewn from top to bottom with parallel rows of stitching to form the five pockets, the central pocket being the widest. In appearance the filled bag is similar to a cricketers leg pad and when required for use is assembled around one side of Sonderkart 1 in a similar manner and inserted in the case. This section has no igniter.

Section 3 (Haupkart) is approximately 26 inches in length and consists of a bundle of tubular cord propellant assembled around a central propellant cylinder and contained in a white cylindrical bag. The central cylinder of propellant extends through the length of the bundle and has an external diameter of 1·4 inches. The internal diameter is 1·2 inches. An igniter containing 80 grams of Nz.N.P. (1,5–1,5) is stitched to the base of the bag.

Section 4 (Vorkart 3) is approximately 27·2 inches in length and consists of a bundle of tubular cord propellant assembled around a central propellant cylinder and contained in a white cylindrical bag. The central cylinder of propellant extends through the length of the bundle and has a large cork disc, in the form of a washer,

placed over its front end which protrudes from the bundle. The external and internal diameters of the cylinder, which accommodates Section 5 (Vorkart 4) when the latter is used, are 2·75 inches and 2·5 inches respectively. The bag has an igniter containing approximately 40 grains of Nz. Man.N.P.(1,5–1,5). The forward part is marked " VORN " to indicate that the section should be loaded with this end to the front.

Section 5 (Vorkart 4) is approximately 26 inches in length and consists of a small bundle of tubular cord propellant contained in a white cylindrical bag of comparatively small diameter. The choke at the forward end of the bag is marked " 4 ". There is no igniter. This section is inserted into the propellant cylinder in Vorkart 3 when used.

The bag of each section is stencilled in black to indicate the designation of the section and the equipment, the weight, nature and size of the propellant charge, the place and year of manufacture of the propellant and the place, lot, month and year of filling. Sections suitable for use in hot climates are marked in red " AUCH FUR TROPEN " or " FUR TROPEN P.T.+ 25°C.".

Cases

Three types of case are known to be used. These are 28·5 inches in length and taper from 7·8 in front of the flange to 7·4 inches at the mouth. The first is a solid drawn brass case with the model number 6324 stamped in the base. The second type is a steel case of similar construction to the first. This case may be coated with brass or be rustproofed and has the letter " St " after the model number.

The third type is a built-up case of steel and is of unusual design. The case consists of a coiled body which is attached to the base by a retaining plate and screwed collar assembled on the primer boss. The body is formed from a four-sided sheet of steel which is shaped and coiled so that there are three and a quarter turns at the base end and only about one and a quarter at the mouth. This is apparently intended to give the case greater strength at the base. One edge of the coiled sheet forms an inclined overlapping joint extending along the length and partially round the body. A layer of black wax is used between the overlapping coils presumably to assist in waterproofing. At the base end the coiled wall is turned inwards to form a curved internal flange corresponding to the upper side of the steel base. The body is made from low carbon rimming steel and has a V.D. hardness figure increasing, rather irregularly, from 105 near the base to 133 at the mouth.

The steel base has the usual external flange or rim and primer hole. The primer boss is screwthreaded externally to receive the screwed steel collar which bears on the steel retaining plate fittings around the boss and overlapping the internal flange on the body. The retaining plate is circular with a central hole to fit over the primer boss and has two circular grooves near its circumference where it is

curved upwards to correspond with the flange on the body. A cardboard disc with its surface covered with black wax is inserted beneath the retaining plate to seal the joint. The stamping in the base of the case includes the model number " 6324/78C " and the designation " 17 cm. K.Mrs. L.". Four equally spaced circular recesses are formed in the base for the purpose of assembly.

Packing

The case and charge sections are packed as follows :—

Charge Section, etc.	Quantity	Package	Weight
Sonderkart 1 and 2 ..	1 of each	Cylinder	61 lb.
Haupkart in case ..	1	Cylinder	84 lb.
Vorkart 3	1	Cylinder	43 lb.
Vorkart 4	24	Box	134 lb.

GERMAN 20 cm. LIGHT SPIGOT MORTAR H.E. ROUND. (20 cm. Wgr.40)

(Figs. 18 and 19)

The 20 cm. H.E. bomb, Model 40, is fired from the 20 cm. Leichter Ladungswerfer (a spigot mortar with a 9 cm. spigot) with a separately loaded cartridge.

The streamlined bomb has an ogival head with a Wgr.Z.36 fuze, or a plug with a lifting loop, at the nose and has a tail tube carrying six vanes. The exterior is painted the normal deep olive green and stencilled in black. The stencilling includes the H.E. numeral " 13 " near the nose (indicating amatol) and the weight class on the cylindrical part of the body. A fuzed bomb bearing the weight class marking " N " weighed 48·75 lb. The overall length, including the fuze, was 31·15 inches and the maximum diameter, approximately 20 cm. The bomb is supplied plugged and fitted with a cylindrical cover of cardboard and steel for the protection of the tail vanes.

The cartridge is short and cylindrical, the lower part being of steel with a partially flanged adapter for the C/23 electric primer at the base and a bakelite upper part. The approximate dimensions are : length 2·2 inches, diameter 3·5 inches. A label giving details of the igniter and propellant charges is affixed to the top of the cartridge.

Bomb

The bomb body is of pearlitic malleable cast iron with a screw-threaded fuze hole at the nose and a tubular extension formed at

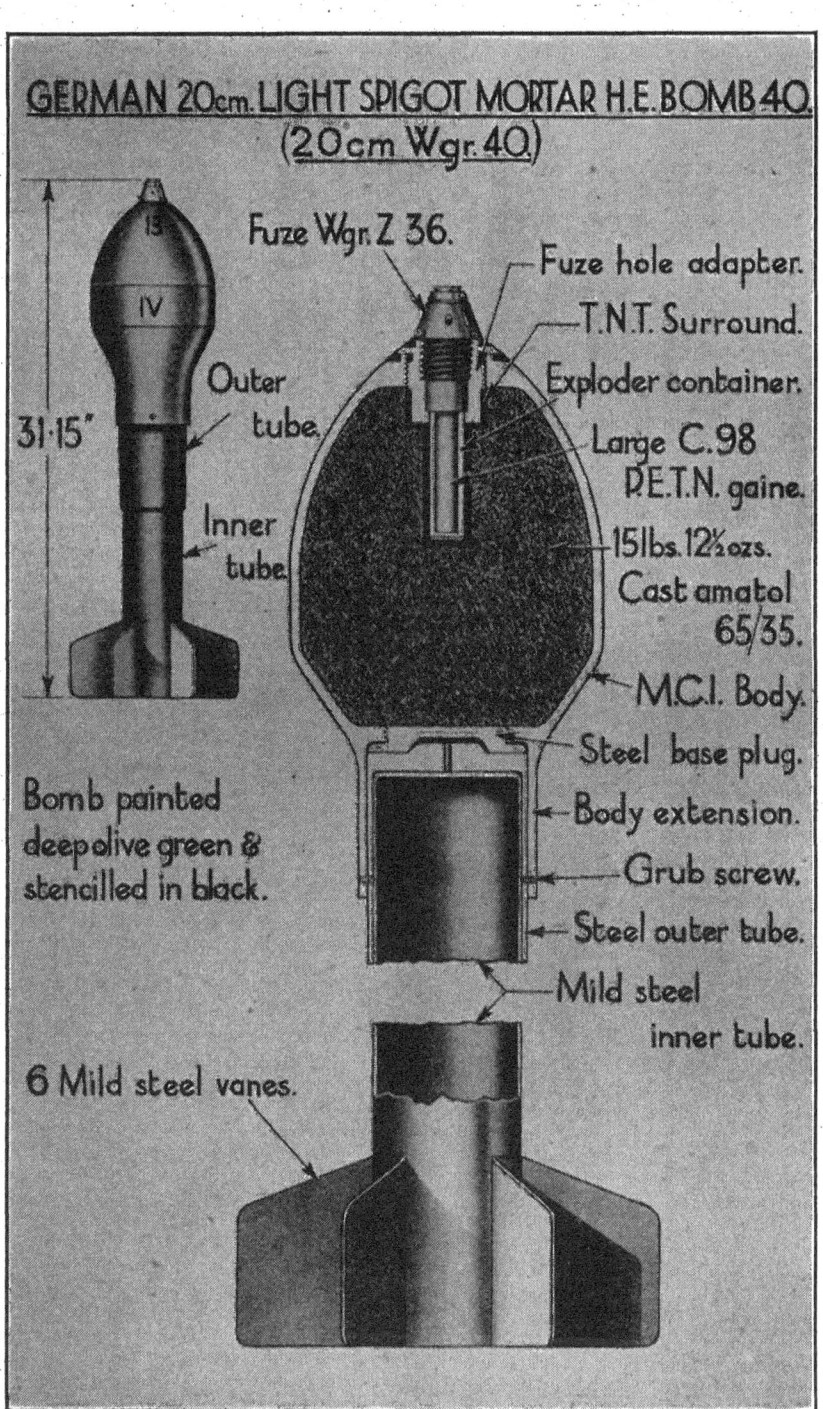

Fig. 18

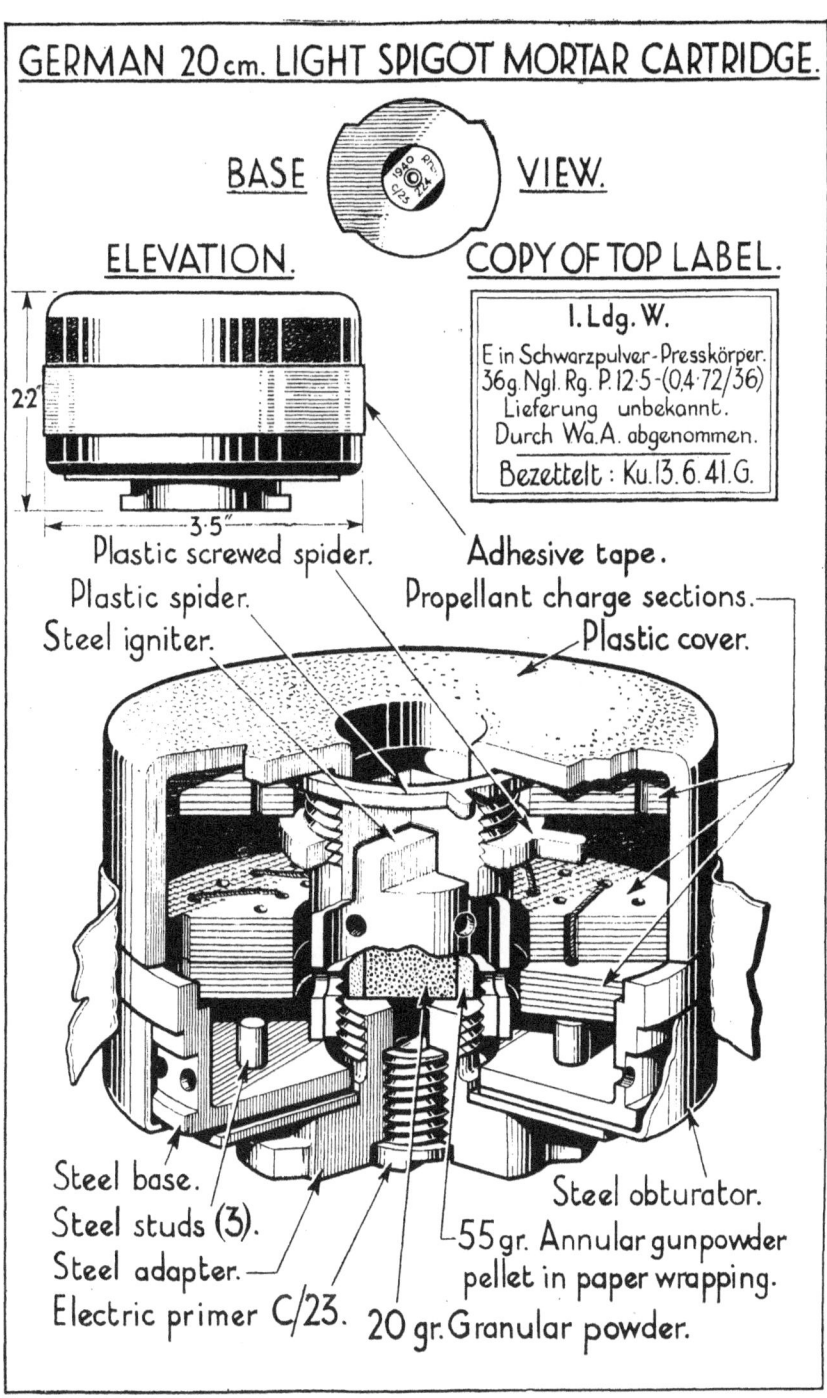

Fig. 19

the base for the attachment of the tail unit. A large hole in the base of the body, within the tubular extension, is closed by a steel screwed plug. A steel fuze hole adapter, inserted at the nose, carries an exploder container of mild steel.

The tail unit consists of a short outer tube and an inner tube which carries the vanes. The outer tube is of chromium-silicon steel with a V.D. hardness figure of 262 and is closed at the front end where the thickness is increased to correspond to a recess in the base plug. A small central hole in the closed end is formed probably for the escape of air during assembly. The inner tube is of mild steel and is also closed at the front end. The two tubes are retained in the tubular extension of the body by two grub screws inserted at diametrically opposite positions in the extension. The rear part of the inner tube protrudes from the inner tube and has six vanes welded to it in pairs.

The weight of the empty bomb is 30·97 lb., the body alone weighing 16·65-lb.

Method of Filling

The bursting charge consists 15-lb. $12\frac{1}{2}$-oz. of cast amatol 65/35 with a thin surround of T.N.T. to the exploder cavity. The exploder container carries the larger size of the C/98 P.E.T.N. gaine.

Fuze and Gaine

The Wgr. Z.36 fuze is described in this pamphlet as a separate item.

The gaine " Gr.Zdlg.C/98 Np " is described in Pamphlet No. 6.

Cartridge

The cartridge is in the form of a short cylindrical box consisting of a steel cup-shaped body with a cover of moulded plastic. The propellant charge, in three sections, is contained inside with a steel igniter containing gunpowder.

The steel body has a hole in the base for the assembly of an adapter which carries the primer and around its exterior has a groove which is connected to the interior by a ring of radial holes and is covered by a steel obturating cup fitted over the base. The cup is expanded by the pressure of propellant gases through the holes. Inside the body there are three equally spaced steel studs protruding from the base for the support of the lowest section of the propellant charge. The obturating cup is supported at the base by a steel disc and the adapter. The steel adapter has an interrupted flange to enable the cartridge to be inserted in the top of the spigot and to be locked to the spigot by turning. A screwthreaded primer hole is formed in its base and at the front end it is screw-threaded externally to receive the steel igniter which secures it to the body.

The steel body of the igniter is in the form of a perforated cylinder

which is closed at the top where it is shaped to form two flat surfaces for the tool used in screwing it to the front end of the adapter.

The plastic cover is an inverted cup with a cylindrical centre piece formed inside which is shaped to fit over and surround the igniter and has corresponding perforations. The centre piece is screwthreaded near the top to receive a screwed plastic spider which supports one of the charge sections. A second spider without a screwthread is used as a distance piece between the top of the charge section and the cover. The top of the cover is recessed and carries a white paper label giving the particulars of the propellant.

Both body and cover are stepped so that the cover fits into the body and the junction is sealed with a wrapping of adhesive tape.

Method of Filling

The igniter contains a 55 grain, paper wrapped, annular pellet of gunpowder inside of which there is a 20 grain filling of small grain gunpowder.

The 36 gram propellant charge of " Ngl.Rg.P.–12·5–(O,4·72/36)," a double base nitroglycerine propellant in ring form, is divided into three sections of equal weight. Each section in the cartridge examined weighed 186 grains and consisted of a number of annular discs perforated with two rings of holes and roughened by an impressed pattern on both sides. The discs are secured by silk ties threaded through the perforations in three places. Each of the discs is between 0·014 and 0·018 inch thick and has a diameter of 2·8 inches. The diameter of the central hole is 1·4 inches. One section of the charge, with another placed on top of it, is supported on the three studs inside the steel body. The third section is carried inside the plastic cover where it is separated from the cover by a spider of plastic and supported by a similar spider screwed to the centre piece.

Primer

The electric primer C/23 is described in this pamphlet.

www.ingramcontent.com/pod-product-compliance
Lightning Source LLC
Chambersburg PA
CBHW032010080426
42735CB00007B/563